Nanostructured Electrochromic Materials for Smart Switchable Windows

Nanostructured Electrochromic Materials for Smart Switchable Windows

Avinash Balakrishnan
Praveen Pattathil

CRC Press
Taylor & Francis Group
Boca Raton London New York

CRC Press is an imprint of the
Taylor & Francis Group, an **informa** business

CRC Press
Taylor & Francis Group
6000 Broken Sound Parkway NW, Suite 300
Boca Raton, FL 33487-2742

First issued in paperback 2019

ISBN-13: 978-1-138-36157-7 (hbk)
ISBN-13: 978-0-367-60664-0 (pbk)

Library of Congress Cataloging-in-Publication Data

Names: Balakrishnan, Avinash, author. | Pattathil, Praveen, author.
Title: Nanostructured electrochromic materials for smart switchable windows / by Avinash Balakrishnan and Praveen Pattathil.
Description: Boca Raton, FL : CRC Press/Taylor & Francis Group, 2018. | Includes bibliographical references and index.
Identifiers: LCCN 2018039858| ISBN 9781138361577 (hardback : acid-free paper) | ISBN 9780429432644 (ebook)
Subjects: LCSH: Windows--Materials. | Glass, Photosensitive. | Electrochromic devices--Materials. | Nanostructured materials--Optical properties.
Classification: LCC TP862 .B35 2018 | DDC 690/.1823--dc23
LC record available at https://lccn.loc.gov/2018039858

Visit the Taylor & Francis Web site at
http://www.taylorandfrancis.com

and the CRC Press Web site at
http://www.crcpress.com

Contents

Preface

THE AIM OF COMPILING this book has been to give a working knowledge of the important details of materials used in electrochromic devices, their characteristics, the process of manufacture, and uses in the industry to all researchers, scientists, and engineering students. This book is structured in a systematic manner and easy-to-read style while presenting fundamentals that can be quickly grasped by a beginner. We believe that it will serve as a comprehensive foundation and reference work for future studies within the rapidly expanding field of electrochromic materials and devices.

Electrochromic (EC) smart windows offer a well-designed approach to dynamically regulate daylight and solar radiation in multiple applications. Electrochromic materials can alter their properties, i.e., change their light transmittance state under the effect of a small electrical voltage or current. Different classes of materials such as transition metal oxides, metal-coordinated complexes, conjugated polymers, and organic compounds show this reversible switching mechanism, which make EC materials eligible for applications such as low-energy consumption displays, light-adapting mirrors in the automobile industry, and smart windows for which the amount of transmitted light and heat needs to be regulated.

The first part of this book describes the different classes, properties, applications, and emerging technologies of smart windows. The second part highlights the EC configuration, different materials used for the fabrication, the role of nanostructure materials, and device fabrication technologies, and the third part focuses on the advancements in electrochromic technology and discusses the challenges associated with electrolytes. The last section of this part rounds off this book by elaborating on the challenges faced in packaging of smart windows.

Avinash Balakrishnan
Manager, Materials Laboratory,
Suzlon Blade Technology Center,
Bhuj, India

Praveen Pattathil
Assistant Manager, Composite Materials,
Suzlon Advanced Technology Center,
Mumbai, India

Authors

Dr. Avinash Balakrishnan joined Suzlon Energy Limited as Manager in 2016, where he heads the Materials Laboratory for Suzlon's Blade Technology vertical. He received his PhD and MS in Materials Engineering from Paichai University, South Korea. He is an alumnus of the National Institute of Technology, Karnataka, where he completed his bachelor in Metallurgical Engineering. His professional experience includes being a research scientist at Korea Research Institute of Standards and Science (KRISS), South Korea, where he worked extensively on ceramic materials for structural and high temperature applications. He was a post-doctoral fellow at Grenoble Institute of Technology (Grenoble-INP), France. He also headed the R&D division for English Indian Clay Limited, India, from 2015 to 2016. His current research interests include developing nanostructured materials for a variety of applications including energy storage, composites, and ceramics. He has co-authored more than 85 research publications, 2 books, and filed 1 patent.

Dr. Praveen Pattathil is an Assistant Manager at the Suzlon Advanced Technology Centre (SATC), India, specializing in nanomaterials for wind turbine blade applications. He received his MS in Physics from Bharathiar University, India, and his doctoral degree in Physics and Nanoscience from the University of Salento, Italy. His doctoral work focused on smart windows, in particular plasmonic nanocrystal synthesis, and the development methods for their integration into novel nanomaterials with systematic investigation of their optical and electronic properties for electrochromic window applications. Dr. Praveen's current research at SATC focuses on the development of light weight nanocomposite materials for wind energy application.

Dr. Praveen has received several recognitions in his research career including the Best High Impact Factor Paper Award from the Italian Institute of Technology, Young Research Fellowship from the University of Salento, and MNRE fellowship from The Ministry of New and Renewable Energy, India. He has produced more than 13 research publications, given 5 conference presentations, and authored 1 book chapter. He is a member of the Materials Research Society, the American Chemical Society, and the Electrochemical Society.

CHAPTER 1

Types, Properties, Applications, and Emerging Technologies of Smart Windows

1.1 OUTLINE

Windows can have a substantial effect on a building's interior design, operating costs, health, productivity, and resident well-being. The purpose of old-fashioned windows was to provide sunlight, view, and fresh air for the residents. However, these days, commercial buildings have become completely sealed, air-conditioned, and electrically lit [1–3]. For this reason, energy consumption has dramatically increased in modern buildings over the past few years. According to the annual report presented by the U.S. Energy Information Administration (EIA), buildings consume nearly 47.6% of the total energy production and emit

38% of the CO_2 into the atmosphere [4], resulting in global warming [5]. Thus, windows are the most important components influencing the building's energy use and highest electricity demand. Approximately 60% of all the energy used in the building sector is consumed for lighting, space heating, and cooling [6]. It is anticipated that this demand for energy will rise by more than 30% by 2020 [7,8].

Improvements in building envelopes and ventilation can play an important role in reducing space heating and cooling consumption levels [9]. Generally, windows can reduce electric lighting loads by allowing natural light to pass through. A proper window design can also cut peak electricity and cooling loads by restricting the heat flow from outside. This can not only reduce the operation cost of the cooling system but also downsize the indoor cooling requirement, thereby saving capital costs. This has limited the role of traditional windows in addressing residents' needs. However, now the trend is shifting from human-centered designs toward improvements in building energy performance.

Generally, window design is influenced by two factors: (a) the climate and (b) the building type (or space configuration within a building). Once these factors are considered, the orientation of the window, its size and type, and required shading systems are addressed [10]. Designers also explore the wide range of possible window materials and assemblies based on the mechanisms of heat transfer. To achieve an energy-efficient window system, the designer generally adopts and/or combines glazing, glass compositions, assembly techniques, and smart technologies.

There are five properties of windows that are the basis for quantifying energy performance [11–14]:

1. U-factor

2. Solar Heat Gain Coefficient (SHGC) or shading coefficient (SC)

3. Visible Transmittance (VT)

4. Air Leakage

5. Light-to-Solar-Gain (LSG) ratio

U-factor: When there is a difference in temperature between indoor and outdoor, the heat is lost or gained through the window frame and glazing due to the combined effects of conduction, convection, and long-wave radiation. The U-factor of a window characterizes its overall heat transfer rate or insulating value. The higher the U-factor, the more heat is transferred (lost) through the window in winter. The unit of U-value is Watt per square meter (W/m^2). U-factors usually range between 1.3 (for a typical aluminum frame single-glazed window) to 0.2 (for a multi-paned, high-performance window with low-emissivity coatings and insulated frames) [15,16]. A window with a U-factor of 0.6 will lose twice as much heat under the same conditions as one with 0.3.

Solar Heat Gain Coefficient (SHGC): Regardless of the outdoor temperature, heat can flow through windows by direct or indirect solar radiation. The ability to regulate this heat gain through windows is characterized in terms of the solar heat gain coefficient (SHGC) or shading coefficient (SC) of the window. The increment in the SHGC ratio is an indication of the increment in the solar gain potential through a given window, and normally the ratio is between 0.0 and 1.0. When SHGC becomes zero, it indicates that none of the incident solar gain is transmitted through the window as heat. All incident solar energy is transmitted through the window as heat when SHGC equals 1.0 [15–17]. For example, a window with an SHGC of 0.6 will admit twice as much solar heat gain as one with 0.3 [18]. Hence, windows with high SHGC values are desirable in buildings where passive solar heating is needed, and low SHGC values are desirable when the loads due to air conditioning is high. The term "SHGC" is rather new and is intended to replace the term "shading coefficient (SC)." The SC of a glass is defined as the ratio of the solar heat gain through a given glazing as compared to that of clear, 1/8 inch single-pane glass [18].

Visible Transmittance (VT): This is an optical property that indicates the amount of visible light transmitted through the glass. It affects energy by providing daylight that creates the opportunity to reduce electric lighting and its associated cooling loads. Sunlight is composed of a range of electromagnetic wavelengths, which are categorized as ultraviolet (UV), visible, and infrared (IR) regions and are collectively referred to as the solar spectrum. The short, UV wavelengths are invisible to the naked eye and are responsible for color fading of fabric and skin damage. Sunlight is made up of 47% of the visible light and 46% of longer IR wavelengths [17,18]. For a given glazing system, the term "Coolness index (K_e)," (also called efficacy factor), is determined by the ratio of the VT to the SC.

Air Leakage: This refers to the amount of air passing through a unit area of window under given pressure conditions. Air leakage can also lead to heat loss and gain through cracks around the frames of the window assembly [18,19].

Light-to-Solar-Gain (LSG) ratio: This is a ratio of VT and SHGC which has been standardized within the glazing industry to allow accurate comparison of windows [18].

$$LSG = \frac{VT}{SHGC} \tag{1.1}$$

A higher LSG ratio means sunlight entering the room is more efficient for daylighting, especially for summer conditions where more light is desired with less solar gain. This ratio is the measurement used to determine whether the glazing is "spectrally selective." In absolute percentage terms, a ratio greater than 1 signifies that the daylight passing through the glass is more than the sun's direct heat passing through it. Selective glasses on the market today have a visible transmittance between 34.0% and 69.0% and a solar heat gain coefficient between 24.0% and 56.0%, with a selectivity index LSG between 1.28 and 2.29 (see Table 1.1) [18].

Window designs should comply with the International Energy Conservation Code (IECC) and American Society of Heating, Refrigerating, and Air-Conditioning Engineers (ASHRAE)

TABLE 1.1   Commercially Available Glazings with Specification

No.	Company Name	VT (%)	SHGC	LSG	U (W/m²)
1	Pilkington Suncool	60	0.32	1.87	1.00
2	AGC Stopray Ultra	60	0.28	2.14	1.00
3	SGG Cool-Lite Expreme	60	0.28	2.14	1.00
4	Guardian Sunguard SNX	62	0.27	2.29	1.36

Source: Lee, E.S., and D.L. DiBartolomeo, *Solar Energy Materials and Solar Cells*, 2002. 71(4): p. 465–491. [12]; Jennifer, S. et al., *Tips for Daylighting with Windows: The Integrated Approach*, 2nd Edition. 2013. [16]; Armistcad, W.H., and S.D. Stookey, *Phototropic material and article made therefrom*. 1965. [43]

Standard 90.1 with energy efficiency provisions for new systems and equipment in existing buildings [20,21]. Both these standards have been developed to provide minimum energy efficiency baselines based on other standards (examples are ASHRAE Standard 189 and the International Green Construction Code [IGCC]) with more ambitious energy efficiency and sustainable design requirements. These design standards provide baselines for green construction that can be adopted and enforced by jurisdictions. For all these standards, certified fenestration energy ratings are required, which must be audited by independent laboratories in accordance with National Fenestration Ratings Council (NFRC) standards [22–24].

Smart windows are energy efficient and comply with environmental building standards. Various coatings, tints, and glass surface treatments are employed to tune the energy properties of windows. To obtain an energy-efficient insulating glass unit (IGU), gas is filled between the glass panes, thus improving heat transfer by lowering the U-factors [25,26]. Switchable, electrochromic glazing can dynamically change their properties to regulate sunlight, heat, glare, and view [27,28]. Window-integrated solar collectors can not only generate energy but also form the integral part of the building envelope [29]. Dynamic window technologies target both commercial and residential markets including new construction and accessory. Approximately 400 million square feet of windows are installed annually in the United States alone. This installation number represents <10% of the annual global demand [30], which indicates that there is a large potential market for smart windows.

The subsequent sections provide a general overview of the types of smart windows that are currently available on the market. They also discuss their properties and their potential for daylight and heat control in buildings.

1.1.1 Smart Windows

Smart windows or switchable windows are comprised of glass or glazing whose light transmission properties are altered when voltage, light, or heat is applied [31–34]. Smart windows can control the throughput of visible light and heat flow into buildings. This can impart energy efficiency by having different transmittance levels depending on the requirements. Unlike shades, smart windows have the ability to partially block the light while maintaining a clear vision of what lies behind the window. Smart glass can be precisely tuned to control the intensity of light, glare, and heat passing through a window. These days, in both commercial and household buildings, facades with smart glass technology are utilized, which reduces the dependency on air conditioning during the summer months and heating during the winter. The capability of switchable glass to provide controllable light conditions during peak and off-peak time of the day makes it valuable and unique. Expectations of demand growth for switchable glass are high. Besides their increasing installation in green energy efficient buildings, demand for smart switchable glass windows is also expected to grow in the automotive, marine, power generation, and construction sectors. The global smart glass market size is targeted to reach $9.98 billion by 2025 [35,36].

Glazing in both commercial and residential segments is expected to grow and have positive impact on the market over the next few years. Architects are recommending the use of smart window technology for new buildings, which is expected to increase smart glass demand. Growth opportunities also exist in the transportation sectors as the demand for smart glass to substitute conventional glass is slowly increasing [37].

The global market for smart windows comprises mainly countries from North America, Asia Pacific, and Europe. Among these regions, North America has emerged as the dominant market in 2014, with a share of 31.8%, which is projected to rise at a compound annual growth rate (CAGR) of 14.7% between 2018 and 2021 [35,38,39]. This is due to rapid technological advancements, high demand for energy-efficient products, and supportive government regulations. Inspired by these factors, the installation of smart glass technology has significantly increased across the automotive, construction, marine, and other sectors in the region. Europe held the second-largest market share in 2014 and is expected to exhibit a CAGR of 15.8% in the next three years [40,41]. The demand for smart window technology from Asia Pacific is also expected to rise in the upcoming years. Rapid infrastructural development and urbanization witnessed across emerging economies like India and China are fueling the demand for energy in the region, which will bolster sales prospects for smart glass manufacturers in the upcoming year. Research Frontiers Inc, Saint-Gobain S.A., View, Inc., Asahi Glass Co., Ltd., and SAGE Electrochromic, etc. are going to be some of the key global players in the market [42]. Currently, a number of different glass technologies are available in the market, such as photochromic, thermochromic, liquid crystal display, and electrochromic. The following section briefly introduces each of these technologies.

1.1.2 Types of Glass Technologies for Smart Windows

Based on the mode of operation, smart windows fall into two main categories (see Figure 1.1): passive control (self-regulating) and active control (adjustable to user's preferences).

1.1.2.1 Passive Windows

Passive dynamic systems do not require an electrical stimulus for their operation. These systems respond independently to the

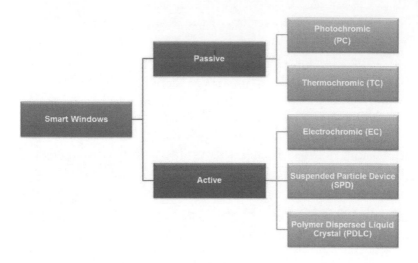

FIGURE 1.1   Classification of smart windows.

presence of natural stimuli such as light (photochromic glass) or heat (thermochromic and thermotropic glazing). Compared with active systems they are easier to install and more reliable in the face of the impossibility of being controlled by the user on request.

1.1.2.1.1 Photochromic (PC)   Photochromic glass was introduced in the early 1960s. It was invented by William H. Armistead and Stanley Donald Stookey of Corning Glass Works [43]. Conventional photochromic glass contains silver-based crystals and performs like an old-fashioned photographic film which darkens (because silver crystals agglomerate into microscopic bits of silver) when light falls on it [44,45]. Only trace quantities of silver crystals are included in the glass composition (<0.1% by volume), and the diameter of each crystal is kept to <0.1 μm. Depending on the field of application, PC materials can be manufactured as liquid solutions, polymer films, amorphous or polycrystalline thin films on flexible or stiff substrates, silicate or polymer glasses, or single crystals [46]. The most widely used PC materials are polymeric materials which are based on organic compounds, such as spiropyrans and dithizonates of metals; activated crystals of alkali metal-halide compounds, e.g., KCl, KBr, or NaF; silicate PC glasses containing silver halide microcrystals, e.g., AgBr or AgCl; and doped salts or oxides of alkaline-earth metals, such as CaF_2/La, Ce and $SrTiO_3$/Fe, Mo [47–51]. Modern photochromic glasses are generally made up of transparent polymers, and instead of silver they contain organic molecules called naphthopyrans that react to light in a slightly different manner [52,53]. They subtly change their molecular structure when UV light is incident on them. The absorption spectrum of these transparent polymers is different than the glasses as they absorb more light and darken significantly. They emulate the opacity of blinds on a sunny day as they progressively block out more light and return to their transparent state in absence of light.

One of the most commonly reported limitations of PC glasses is that their switching time from darkened state to transparent state is longer. For instance, PC glasses exposed to sunlight will allow only about 60% of light to pass through them for the first 5 minutes [54,55] and will take an additional hour to achieve complete transparency. Another major disadvantage is that the PC compounds depend on the presence of heat and light to undergo optical changes. This makes PC glasses more effective in winter and less so during summer. Studies have shown that PC glasses are less effective inside cars and other automobiles as the glass windows of these vehicles block UV rays. Thus, the PC molecules do not receive enough UV rays to undergo changes and be optically effective [56]. Notwithstanding these few limitations, PC glasses present a brilliant piece of optical engineering and are convenient for applications where switching speed is not a critical requirement.

1.1.2.1.2 Thermochromic (TC) The word "thermochromic" originates from the Greek: "thermos" meaning warm or hot; and "chroma" meaning color. TC materials can modulate their color in response to temperature variations [57]. This type of glass can reduce the unwanted solar energy gain by changing the device's reflectance and transmission properties. The TC thin films at room temperature exist in a monoclinic state [58] and behave like a semiconductor. They show less reflectiveness especially in near-IR (NIR) radiation. With the increase in temperature, TC material transitions from monoclinic to rutile phases. This transition is called metal to semiconductor transition (MST). In the rutile phase (high temperature state) TC materials act like a semi-metal and reflect a wide range of solar radiation. At high temperatures, TC windows reduce NIR and far-IR transmittance. While at low temperatures, they allow these regions of solar radiation to pass through. The MST is a reversible phenomenon seen with large variations in electrical and optical properties in NIR range [59,60].

The fabrication and glazing properties of TC materials are not new in the glass industry and have been widely studied [61,62]. The most promising TC material for windows is VO_2 which is known to exist in four polymorphic forms: monoclinic VO_2 (M) and rutile VO_2 (R) and two metastable forms VO_2 (A) and VO_2 (B) occurring during the monoclinic to rutile phase transition at a temperature of 68°C [63]. However, VO_2-based TC windows suffer from low luminous transmittance at room temperature (between 40% and 50%, which is well below the acceptable value of 60%). This limitation could be solved by fluorination [64] or applying SiO_2 anti-reflective (AR) coating [65,66]. ZrO_2 coating with an appropriate refractive index can also enhance the luminous transmittance while retaining the TC switching [67,68]. There are some compounds which can be used as precursors for VO_2 deposition. The most common precursors are vanadium alkoxides [69,70] and oxovanadium reagents [71,72]. Vanadium chloride should react with a source of oxygen (such as methanol or ethanol), the product of which is V_2O_5. Heating the product in a reducing atmosphere results in deposition of VO_2. Addition of tetraoctylammonium bromide (TOAB) as surfactant to the vanadium dioxide matrix can control the distribution, shape, and size of the synthesized particles. It is reported that surfactant TOAB can reduce the transition temperature from 50–54°C to 42–45°C and from 52°C to 34°C [71]. As mentioned in Section 1.1.2.1.2, the MST temperature of pure VO_2 is about 68°C which is relatively a high temperature. In order to make this type of glazing feasible, the MST temperature should decrease to near the room temperature. Doping metal ions into the lattice of TC materials can alter MST temperature. The size and charge of dopant ion, film's strain, as well as the variations in electron carrier density are the factors which determine the MST temperature [73,74]. Besides the disadvantage of low luminance visibility, low energy-saving efficiency also makes the application of VO_2 coatings limited. The change in transmittance before and after MST temperature at 2500 nm is termed as switching efficiency (ηT). ηT is considered as the benchmark of energy-saving efficiency which is influenced by doping, microstructure, and film thickness [74]. The most critical factor among them is the film thickness that significantly affects the ηT. However, increasing the film thickness has an adverse effect on visible transmittance. In order to overcome these limitations certain critical steps have to be taken. These include (a) choosing the most suitable dopant (for reducing MST temperature and improving transmittance), (b) employing appropriate coating technology (to acquire the optimum thickness and sufficient TC transition; the ideal film thickness should be between 40 and 80 nm), (c) adding efficient anti-reflecting coating (to increase visible transmittance), and (d) reducing the coating costs [75].

Transparent insulated glass units incorporating a TC laminate and low-e coating can have a visible light transmission ranging between 54% and 8%, a SHGC from 0.36 and 0.16, and a U-value of 0.24 [10,75,76]. Performance can be modulated by selecting glass tints of different compositions, low-e coatings, and air spaces. TC glasses can be used in operable or fixed windows, doors, and skylights. TC windows can be installed by glazing contractors, just like traditional windows, and they do not require wires, power supplies, or control equipment. These features, along with lower cost, make TC windows a very attractive approach to variable tinting windows. Suntuitive Glass is one of the commercially available TC glasses manufactured from Pleotint, LLC [42]. Units as large as 64″ × 144″ have been produced with the TC interlayer and installed throughout the world. The data on the interlayer used by Suntuitive is available in the International Glazing Database (IGDB), and ratings are available through the National Fenestration Rating Council (NFRC) [23,77,78].

1.1.2.2 Active Windows

Active windows represent a family of electrically-active glass technologies capable of darkening at the press of a button to block light and prevent heat from entering the building. This technology started out as a luxury feature in high-end projects, but studies have pointed to the economic benefit, which is peak electricity savings of about 30% that can be obtained by blocking solar radiation into buildings [79–81]. This can downsize the air conditioning unit or use it at lower power settings. Active windows can be manually controlled or can be connected to a computerized building management system (CBMS) to respond to subtle changes in temperature, artificial and natural lighting levels, heat intakes, and presence of people [82,83]. They can also tune themselves to the needs of users and adjust the intensity of incoming visible light and infrared radiation without the use of screens. Some of the most advanced systems on the market provide integration with photovoltaic systems for total electrical self-sufficiency. In addition, there is the possibility of accessing these systems through smart phones, which can allow independent adjustment of different panels of the same window. Active windows also present an opportunity of introducing real imaging displays with touch screen technology [84,85].

Electrically controllable active systems include electrochromic (EC) glass, suspended particle devices (SPDs), and liquid crystal devices (LC/PDLC). The most recent genre of active windows is based on micro-blinds (MEMS) and nanocoatings which are still at the experimental stage. Each of these technologies is unique in their characteristics, performance, and costs, making it suitable for customized applications or requirements [86–88].

1.1.2.2.1 Electrochromic (EC) An electrochromic device is composed of an ion conductor (or electrolyte) that is sandwiched between two transparent conductive layers, namely an electrochromic film (also called an electrode) and an accumulation layer (known as counter electrode) [89,90]. When an electric potential difference is applied between the two transparent conductors, ions are released from the accumulation layer, pass through the conductor layer, and are lead into the electrochromic layer, thereby changing its optical properties. This process becomes reversible when the electrical stimulus is turned off, as the ions are released from the electrochromic layer through the conductive layer and are deposited into the accumulation layer making the device transparent again. Glazings are typically green or blue in color in relation to the electrochromic materials. For instance, the tungsten oxide can vary its color from transparent to blue where the degree of transparency can be modulated in intermediate states from clear (device off) to completely tinted. Here, the light transmission varies from approximately 60% in the translucent state to about 1% when opaque with varying SHGC between 0.46 and 0.06. The amount of energy required by the EC is 1–2.5 Wp/m^2 to switch between the different coloration states [91,92]. Due to the unique property of EC materials to sustain a bi-stable configuration, the amount of energy (<0.4 W/m^2) required to maintain the preferred tinted state is even less. During the tinting, darkening slowly starts from the edges, moving inwards. If the device is working properly, the change of properties of the glass over the entire surface is uniform. This process could take several seconds to some minutes depending on panel size and the glass temperature. The tinting process generally takes a little longer than the clearing process. For example, in a tropical climate, a 35 × 60-inch window usually takes around 10 minutes to achieve approximately 90% of its tinting cycle. The time increases in colder conditions when the need to control the tinting of the glass is not as relevant. The subtle changes in light transmission are advantageous because they allow the residents to adapt naturally without ailment or distraction.

However, it should be noted that the EC glass provides visibility even in the tinted state, thereby maintaining visible contact with the outside environment. Some of the commercially sold EC glasses are View and Sage. There is still technological room for improvement of EC glass, where it would be possible to increase the number of control states (currently four), the switching speed, and the opacity in tinted state to ensure privacy and further reduce energy consumption [12,16].

1.1.2.2.2 Suspended Particle Devices (SPDs)   SPDs (commercially available under the brand names of Isoclima, Vision Systems, Innovative Glass, Hitachi Chemicals, etc.) are composed of a double sheet of glass within which a layer of thin laminate of suspended particles like rods is immersed in a fluid. This set-up is placed between two electrical conductors made of transparent thin plastic films. When the power is switched on, the suspended rod particles align themselves in a fashion that allows the light to pass through and the SPD smart glass panel to become transparent [12,16].

When the power is switched off, the suspended rod particles get randomly oriented, thereby blocking the light and making the glass appear dark (tinting to blue or, in more recent developments, gray or black). In this manner, SPD glass can become transparent or opaque, allowing immediate control over the amount of light and heat passing through. SPD smart glass, when completely tinted, can block up to 99.4% of the visible radiation and protect from UV rays both in switched on or off condition. A constant supply of electricity must be provided for SPDs to maintain their particle alignment. Once the electrical supply is interrupted, the particles disperse again in the solution and the glass becomes opaque. In this way, SPDs differ from EC glass, which works under the same principles but requires only a single burst of electricity to transit from darkened to the transparent state.

The typical ranges of VT and SHGC in SPDs are 65%–0.5% and 0.57–0.06 respectively with switching times of a few seconds [93–95]. Attributes like high switching speed and total user controllability makes SPD glass suitable for automobiles (side and rear windows, transparent sunroofs), ships (windows, skylights, portholes, partitions, and doors), and aircrafts (more than 30 different models of aircrafts have installed SPD windows). SPDs require alternating current (AC) with about 100 V to operate from the tinted state to the transparent one and can be modulated to any intermediate translucent states. The power requirements are 5 and 0.55 W/m^2 for switching and to maintain a state of constant transmission [95]. It is anticipated that with new developments, the operating voltages may further drop to approximately 35 V. New material suspensions are also being researched to obtain different colors than blue (i.e., green, red, and purple) to obtain a greater variation in the U-factor. Panels can be sized up to 60 × 120 inches and are available in different shapes (i.e., both planar and curved). Durability, aging, and optical-solar properties are yet to be studied as these products are now entering the market. But the high cost remains a problem (SPDs are the most expensive dynamic glass existing in the market) [5].

1.1.2.2.3 Polymer Dispersed Liquid Crystals (PDLCs) PDLC devices (sold under brand names like Scienstry, Polytronix, Essex Safety Glass, Switch Glass, SmartGlass International, Magic Glass, Dream Glass, etc.) are composed of a polymer matrix film sandwiched between two electrical conductors of transparent thin plastic film [96,97]. The polymer matrix film is dispersed with liquid crystal spheres with a diameter on the same order of magnitude as the wavelength of visible radiation. In the absence of external electrical stimulus, the liquid crystals have a random arrangement and the light rays undergo disordered diffractions. As a result, the glazed elements appear white and translucent. On the other hand, when an electric field is applied, the liquid crystals orient themselves in identical direction ensuring the transparency of the panels. The degree of transparency can be tuned by the voltage applied.

Scattering PDLCs are commercially available and are used as privacy screens or windows. These PDLCs can switch between a transparent/clear state and an opaque/scattering state. Selective chemical dyes can be introduced into the PDLC mixtures, which can scatter red, green, or blue light depending on the preference or requirement. The light transmittance of PDLCs does not normally exceed 70%, while in the switched off state it is about 50%. Appropriate dyes can be added to darken the device in the off state. The main disadvantage of PDLCs is that it does not provide enough block to obtain a significant reduction of the U-factor which is usually between 0.69 and 0.55 [98–100].

If the polymer concentration within a PDLC device is increased to values of 60%–80% and is instantly cured using high intensity UV, PDLCs with smaller droplet sizes can be obtained. Once droplet size reduces to nanoscale, the light passing through the mixture is no longer scattered at optical wavelengths. On application of an electric field, the nano-PDLC mixture can still switch between randomly aligned and vertically aligned states, but no apparent scattering occurs. The transmitted light only gets modulated, which can be determined by the average orientation (and average refractive index) of the LC. Nano-PDLC phase-modulation devices can switch at very high speeds (microseconds), making them appropriate for wave front correction devices in adaptive optics employed in astronomy, line-of-sight communications, and ophthalmic applications [101,102].

PDLC systems find their application in the construction of interior or exterior partitions that require privacy, such as shops, meeting rooms, intensive care areas, bathroom and shower doors, or transparent walls to use as temporary projection screens. One of the earliest and most famous architectural uses was in OMA and IDEO's collaboration on the Prada flagship store in New York City, where among other highly innovative technological applications, PDLC windows were incorporated in the fitting rooms. Furthermore, compared to EC systems, PDLCs are not bistable and require a constantly applied electric field (under

AC conditions) for correct operation, resulting in a continuous consumption of electrical energy (about 5–10 W/m^2 between 65–110 V) [103,104].

1.1.3 Emerging Technologies

Among the different emerging smart window technologies, EC smart windows with integrated micro-blinds and nanocrystals are of great interest for future applications in Architecture. Micro-blinds can regulate the amount of light passing through in response to applied voltage [105]. They are composed of rolled thin metal blinds on glass which are microscopic in nature. These metal layers are magnetron sputtered and patterned by laser or lithography process. The glass substrate is coated with a thin layer of a transparent conductive oxide layer [106]. A thin insulator is deposited between the rolled metal layer and the transparent conductive oxide layer to prevent short circuiting. When no electric field is applied, the micro-blinds are rolled and allow light to pass through the glass. When a voltage is applied between the rolled metal layer and the transparent conductive oxide layer, an electric field is formed between the two electrodes, which causes the rolled micro-blinds to roll out and thus block light [107,108]. The micro-blinds have several benefits, including high switching rates (milliseconds), UV durability, customized transmission, ease of fabrication, and cost-effectiveness. In the quest to develop market sized products, the focus is on the development of laser etching techniques. From the nanotechnology standpoint, indium tin oxide (ITO) nanocrystals have been developed which can be embedded in a glassy matrix of NbO_2. The resulting compound can regulate the incoming visible light and infrared radiation (NIR), thereby eliminating any unwanted energy intakes while exploiting the natural light [109,110].

Utilization of smart windows can bring numerous advantages in terms of energy efficiency, comfort, and architectural quality of buildings. Static windows with selective type of glass cannot provide complete solutions to optimize solar gains and light conditions. This limits the size of glazed components in the design phase. Solutions coupling automated dynamic intelligent windows with building automation systems offer excellent energy performance. However, high installation, maintenance, and management costs along with restricted view from the inside to the outside limit their application. In contrast, EC window technologies can modulate the amount of incoming light and heat according to the requirement, thereby adapting to weather conditions and improving a building's overall performance for every kind of climate. EC systems can significantly reduce the energy consumption for artificial lighting and air conditioning during summer and winter. Automated screen systems can lower installation costs and management. However, the critical barriers in adopting these new technologies still reside in the lack of standardization and limited information between professionals and consumers. Increased market penetration and improvement of manufacturing processes should help acceptance of these upcoming technologies [12,16].

Rather than focusing on improving the already high performance, it is more important to drastically reduce the cost of smart glasses, which means the attention should be on new materials, manufacturing processes, and easier installation. To ensure this, wide market penetration of smart glass should be targeted such that the cost should not exceed 75 $/m^2. This will allow a return on investment time of 10–12 years for residential buildings and 21–22 years for commercial ones. Moreover, for existing buildings, the necessity of having to replace the entire window and the need for extensive wiring for installing smart windows is still an issue. It is therefore important to promote research in the field of dynamic high-performance glazing materials, which can be used to retrofit existing windows. In addition, the development of internal self-powering systems and built-in remote-control systems for windows should be studied [1,37,42].

REFERENCES

1. Granqvist, C.G. Oxide electrochromics: An introduction to devices and materials. *Solar Energy Materials and Solar Cells*, 2012. 99: p. 1–13.
2. Jaksic, N., and C. Salahifar. A feasibility study of electrochromic windows in vehicles. *Solar Energy Materials and Solar Cells*, 2003. 79: p. 409–423.
3. Granqvist, C.G., P.C. Lansåker, N.R. Mlyuka, G.A. Niklasson, and E. Avendaño. Progress in chromogenics: New results for electrochromic and thermochromic materials and devices. *Solar Energy Materials and Solar Cells*, 2009. 93(12): p. 2032–2039.
4. Cannavale, A., F. Fiorito, D. Resta, and G. Gigli. Visual comfort assessment of smart photovoltachromic windows. *Energy and Buildings*, 2013. 65: p. 137–145.
5. Malara, F., A. Cannavale, S. Carallo, and G. Gigli. Smart windows for building integration: A new architecture for photovoltachromic devices. *ACS Applied Materials & Interfaces*, 2014. 6(12): p. 9290–9297.
6. Balaras, C.A., K. Droutsa, E. Dascalaki, and S. Kontoyiannidis. Heating energy consumption and resulting environmental impact of European apartment buildings. *Energy and Buildings*, 2005. 37(5): p. 429–442.
7. Reilly, S., and W. Hawthorne. The impact of windows on residential energy use. *ASHRAE Transactions Atlanta* 1998. 104.
8. Grynning, S., A. Gustavsen, B. Time, and B.P. Jelle. Windows in the buildings of tomorrow: Energy losers or energy gainers? *Energy and Buildings*, 2013. 61: p. 185–192.
9. Badescu, V., and M.D. Staicovici. Renewable energy for passive house heating: Model of the active solar heating system. *Energy and Buildings*, 2006. 38(2): p. 129–141.
10. Baetens, R., B.P. Jelle, and A. Gustavsen. Properties, requirements and possibilities of smart windows for dynamic daylight and solar energy control in buildings: A state-of-the-art review. *Solar Energy Materials and Solar Cells*, 2010. 94(2): p. 87–105.
11. Lampert, C.M. Smart switchable glazing for solar energy and daylight control. *Solar Energy Materials and Solar Cells*, 1998. 52(3): p. 207–221.
12. Lee, E.S., and D.L. DiBartolomeo. Application issues for large-area electrochromic windows in commercial buildings. *Solar Energy Materials and Solar Cells*, 2002. 71(4): p. 465–491.

13. Mathew, J.G.H., S.P. Sapers, M.J. Cumbo, N.A. O'Brien, R.B. Sargent, V.P. Raksha, R.B. Lahaderne, and B.P. Hichwa. Large area electrochromics for architectural applications. *Journal of Non-Crystalline Solids*, 1997. 218: p. 342–346.

14. Ander, G.D. Windows and glazing 2016. https://www.wbdg.org/resources/windows-and-glazing.

15. Carmody, J., S. Selkowitz, E. Lee, and D. Arasteh. *Window Systems for High-performance Buildings*. 2004: Norton.

16. Jennifer, S., E.S. Lee, F.M. Rubinstein, S.E. Selkowitz, and A. Robinson. *Tips for Daylighting with Windows: The Integrated Approach*, 2nd Edition. 2013. https://windows.lbl.gov/publications/tips-daylighting-windows-integrated-approach-2nd-edition.

17. Ghoshal, S., and S. Neogi. Advance glazing system—energy efficiency approach for buildings: A review. *Energy Procedia*, 2014. 54: p. 352–358.

18. Carmody, J., D. Araseth, S. Selkowitz, and L. Heschong. *Residential Windows: A Guide to New Technologies and Energy Performance*. 2000: Norton.

19. Pierucci, A., A. Cannavale, F. Martellotta, and F. Fiorito, Smart windows for carbon neutral buildings: A life cycle approach. *Energy and Buildings*, 2018. 165: p. 160–171.

20. Dussault, J.-M., and L. Gosselin. Office buildings with electrochromic windows: A sensitivity analysis of design parameters on energy performance, and thermal and visual comfort. *Energy and Buildings*, 2017. 153: p. 50–62.

21. Dussault, J.-M., M. Sourbron, and L. Gosselin. Reduced energy consumption and enhanced comfort with smart windows: Comparison between quasi-optimal, predictive and rule-based control strategies. *Energy and Buildings*, 2016. 127: p. 680–691.

22. Chen, F., and S.K. Wittkopf. Summer condition thermal transmittance measurement of fenestration systems using calorimetric hot box. *Energy and Buildings*, 2012. 53: p. 47–56.

23. Chen, X., H. Yang, and L. Lu. A comprehensive review on passive design approaches in green building rating tools. *Renewable and Sustainable Energy Reviews*, 2015. 50: p. 1425–1436.

24. Li, Y., X. Chen, X. Wang, Y. Xu, and P.-H. Chen. A review of studies on green building assessment methods by comparative analysis. *Energy and Buildings*, 2017. 146: p. 152–159.

25. Cui, Y., Y. Ke, C. Liu, Z. Chen, N. Wang, L. Zhang, Y. Zhou, S. Wang, Y. Gao, and Y. Long. *Thermochromic VO$_2$ for Energy-Efficient Smart Windows*. Joule, 2018.

26. Wang, J., and D. Shi. Spectral selective and photothermal nano structured thin films for energy efficient windows. *Applied Energy*, 2017. 208: p. 83–96.

27. Ghosh, A., B. Norton, and T.K. Mallick. Influence of atmospheric clearness on PDLC switchable glazing transmission. *Energy and Buildings*, 2018. 172: p. 257–264.

28. Ghosh, A., B. Norton, and T.K. Mallick. Daylight characteristics of a polymer dispersed liquid crystal switchable glazing. *Solar Energy Materials and Solar Cells*, 2018. 174: p. 572–576.

29. Gautam, K.R., and G.B. Andresen. Performance comparison of building-integrated combined photovoltaic thermal solar collectors (BiPVT) with other building-integrated solar technologies. *Solar Energy*, 2017. 155: p. 93–102.

30. Niklasson, G.A., and C.G. Granqvist. Electrochromics for smart windows: Thin films of tungsten oxide and nickel oxide, and devices based on these. *Journal of Materials Chemistry*, 2007. 17(2): p. 127–156.

31. Granqvist, C.G. Electrochromics for smart windows: Oxide-based thin films and devices. *Thin Solid Films*, 2014. 564: p. 1–38.

32. Casini, M. Smart windows for energy efficiency of buildings. *International Journal of Civil and Structural Engineering*, 2015. 2: 2372–3971.

33. Qian, J., D. Ma, Z. Xu, D. Li, and J. Wang. Electrochromic properties of hydrothermally grown Prussian blue film and device. *Solar Energy Materials and Solar Cells*, 2018. 177: p. 9–14.

34. Assis, L.M.N., R. Leones, J. Kanicki, A. Pawlicka, and M.M. Silva. Prussian blue for electrochromic devices. *Journal of Electroanalytical Chemistry*, 2016. 777: p. 33–39.

35. Pacheco-Torgal, F. Eco-efficient construction and building materials research under the EU Framework Programme Horizon 2020. *Construction and Building Materials*, 2014. 51: p. 151–162.

36. De Oliveira-De Jesus, P.M. and C.H. Antunes. Economic valuation of smart grid investments on electricity markets. *Sustainable Energy, Grids and Networks*, 2018. 16: p. 70–90.

37. Huang, L.-M., C.-W. Hu, H.-C. Liu, C.-Y. Hsu, C.-H. Chen, and K.-C. Ho. Photovoltaic electrochromic device for solar cell module and self-powered smart glass applications. *Solar Energy Materials and Solar Cells*, 2012. 99: p. 154–159.

38. Granqvist, C.G. Smart windows and intelligent glass façades. *Smart Materials Bulletin*, 2002. 2002(10): p. 9–10.

39. Pacheco-Torgal, F., and J.A. Labrincha. The future of construction materials research and the seventh UN Millennium Development

Goal: A few insights. *Construction and Building Materials*, 2013. 40: p. 729–737.

40. Pierucci, A., A. Cannavale, F. Martellotta, and F. Fiorito. Smart windows for carbon neutral buildings. *A Life Cycle Approach*, 2018. 165.

41. Santamouris, M. Innovating to zero the building sector in Europe: Minimising the energy consumption, eradication of the energy poverty and mitigating the local climate change. *Solar Energy*, 2016. 128: p. 61–94.

42. Sibilio, S., A. Rosato, M. Scorpio, G. Iuliano, G. Ciampi, G. Peter Vanoli, and F. Rossi. A review of electrochromic windows for residential applications. *International Journal of Heat and Technology*, 2016. 34: p. 481–488.

43. Armistcad, W.H., and S.D. Stookey. *Phototropic material and article made therefrom*. 1965.

44. Hočevar, M., and U. Opara Krašovec. A photochromic single glass pane. *Solar Energy Materials and Solar Cells*, 2018. 186: p. 111–114.

45. Flores, J.L., G. Garcia-Torales, J.P. Aguayo-Adame, and J.A. Ferrari. Self adaptive high pass filtering using photochromic glass. *Optik - International Journal for Light and Electron Optics*, 2012. 123(12): p. 1067–1070.

46. Cruz, R.P., M. Nalin, S.J.L. Ribeiro, and C. Molina. Photochromic dynamics of organic–inorganic hybrids supported on transparent and flexible recycled PET. *Optical Materials*, 2017. 66: p. 297–301.

47. Pardo, R., M. Zayat, and D. Levy. Photochromic organic-inorganic hybrid materials. *Chemical Society Reviews*, 2011. 40: p. 672–687.

48. Irie, M. Photochromism: Memories and switches introduction. *Chemical Reviews*, 2000. 100(5): p. 1683–1684.

49. Lehn, J.-M. From supramolecular chemistry towards constitutional dynamic chemistry and adaptive chemistry. *Chemical Society Reviews*, 2007. 36(2): p. 151–160.

50. Kawata, S., and Y. Kawata. Three-dimensional optical data storage using photochromic materials. *Chemical Reviews*, 2000. 100(5): p. 1777–1788.

51. Wang, M.-S., G. Xu, Z.-J. Zhang, and G.-C. Guo. Inorganic–organic hybrid photochromic materials. *Chemical Communications*, 2010. 46(3): p. 361–376.

52. Levy, D., S. Einhorn, and D. Avnir. Applications of the sol-gel process for the preparation of photochromic information-recording materials: Synthesis, properties, mechanisms. *Journal of Non-Crystalline Solids*, 1989. 113(2): p. 137–145.

53. Zhang, J., Q. Zou, and H. Tian. Photochromic materials: More than meets the eye. *Advanced Materials*, 2013. 25(3): p. 378–399.
54. Delaire, J.A., and K. Nakatani. Linear and nonlinear optical properties of photochromic molecules and materials. *Chemical Reviews*, 2000. 100(5): p. 1817–1846.
55. Shigeo, T., K. Seiji, and I. Tomiki. Amplified image recording in liquid crystal media by means of photochemically triggered phase transition. *Chemistry Letters*, 1987. 16(5): p. 911–914.
56. Delbaere, S., B. Luccioni-Houze, C. Bochu, Y. Teral, M. Campredon, and G. Vermeersch. Kinetic and structural studies of the photochromic process of 3H-naphthopyrans by UV and NMR spectroscopy. *Journal of the Chemical Society, Perkin Transactions* 1998. 2(5): p. 1153–1158.
57. Gao, Y., H. Luo, Z. Zhang, L. Kang, Z. Chen, J. Du, M. Kanehira, and C. Cao. Nanoceramic VO_2 thermochromic smart glass: A review on progress in solution processing. *Nano Energy*, 2012. 1(2): p. 221–246.
58. Seeboth, A., D. Lötzsch, R. Ruhmann, and O. Muehling. Thermochromic polymers—function by design. *Chemical Reviews*, 2014. 114(5): p. 3037–3068.
59. Lee, M.-H., and J.-S. Cho. Better thermochromic glazing of windows with anti-reflection coating. *Thin Solid Films*, 2000. 365(1): p. 5–6.
60. Morin, F.J. Oxides which show a metal-to-Insulator transition at the Néel temperature. *Physical Review Letters*, 1959. 3(1): p. 34–36.
61. Ryabova, L.A., I.A. Serbinov, and A.S. Darevsky. Preparation and properties of pyrolysis of vanadium oxide films. *Journal of The Electrochemical Society*, 1972. 119(4): p. 427–429.
62. MacChesney, J.B., J.F. Potter, and H.J. Guggenheim. Preparation and properties of vanadium dioxide films. *Journal of The Electrochemical Society*, 1968. 115(1): p. 52–55.
63. Eyert, V. The metal-insulator transitions of VO_2: A band theoretical approach. *Annalen der Physik*, Vol. 11. 2002. 11.
64. Gutarra, A., A. Azens, B. Stjerna, and C.G. Granqvist. Electrochromism of sputtered fluorinated titanium oxide thin films. *Applied Physics Letters*, 1994. 64(13): p. 1604–1606.
65. Lee, M.-H. Thermochromic glazing of windows with better luminous solar transmittance. 2002. 71: p. 537–540.
66. Zhang, W., J. Tu, W. Long, W. Lai, Y. Sheng, and T. Guo. Preparation of SiO_2 anti-reflection coatings by sol-gel method. *Energy Procedia*, 2017. 130: p. 72–76.

67. García-Heras, M., J.M. Rincón, M. Romero, and M.A. Villegas. Indentation properties of ZrO_2–SiO_2 coatings on glass substrates. *Materials Research Bulletin*, 2003. 38(11): p. 1635–1644.

68. Villegas, M.A. Chemical and microstructural characterization of sol–gel coatings in the ZrO_2–SiO_2 system. *Thin Solid Films*, 2001. 382(1): p. 124–132.

69. Nunes, G.G., G.R. Friedermann, J.L.B. dos Santos, M.H. Herbst, N.V. Vugman, P.B. Hitchcock, G.J. Leigh, E.L. Sá, C.J. da Cunha, and J.F. Soares. The first thermochromic vanadium (IV) alkoxide system. *Inorganic Chemistry Communications*, 2005. 8(1): p. 83–88.

70. Seyfouri, M.M., and R. Binions. Sol-gel approaches to thermochromic vanadium dioxide coating for smart glazing application. *Solar Energy Materials and Solar Cells*, 2017. 159: p. 52–65.

71. Kamalisarvestani, M., R. Saidur, S. Mekhilef, and F.S. Javadi. Performance, materials and coating technologies of thermochromic thin films on smart windows. *Renewable and Sustainable Energy Reviews*, 2013. 26: p. 353–364.

72. Blackman, C.S., C. Piccirillo, R. Binions, and I.P. Parkin. Atmospheric pressure chemical vapour deposition of thermochromic tungsten doped vanadium dioxide thin films for use in architectural glazing. *Thin Solid Films*, 2009. 517(16): p. 4565–4570.

73. Sun, J., and G.K. Pribil. Analyzing optical properties of thin vanadium oxide films through semiconductor-to-metal phase transition using spectroscopic ellipsometry. *Applied Surface Science*, 2017. 421: p. 819–823.

74. Lu, Y., and X. Zhou. Synthesis and characterization of nanorod-structured vanadium oxides. *Thin Solid Films*, 2018. 660: p. 180–185.

75. Rezaei, S.D., S. Shannigrahi, and S. Ramakrishna. A review of conventional, advanced, and smart glazing technologies and materials for improving indoor environment. *Solar Energy Materials and Solar Cells*, 2017. 159: p. 26–51.

76. Jelle, B.P., A. Hynd, A. Gustavsen, D. Arasteh, H. Goudey, and R. Hart. Fenestration of today and tomorrow: A state-of-the-art review and future research opportunities. *Solar Energy Materials and Solar Cells*, 2012. 96: p. 1–28.

77. Doan, D.T., A. Ghaffarianhoseini, N. Naismith, T. Zhang, A. Ghaffarianhoseini, and J. Tookey. A critical comparison of green building rating systems. *Building and Environment*, 2017. 123: p. 243–260.

78. Favoino, F., M. Overend, and Q. Jin. The optimal thermo-optical properties and energy saving potential of adaptive glazing technologies. *Applied Energy*, 2015. 156: p. 1–15.
79. Domingues, P., P. Carreira, R. Vieira, and W. Kastner. Building automation systems: Concepts and technology review. *Computer Standards & Interfaces*, 2016. 45: p. 1–12.
80. Biyik, E., M. Araz, A. Hepbasli, M. Shahrestani, R. Yao, L. Shao, E. Essah, A.C. Oliveira, T. del Caño, E. Rico, J.L. Lechón, L. Andrade, A. Mendes, and Y.B. Atlı. A key review of building integrated photovoltaic (BIPV) systems. *Engineering Science and Technology, an International Journal*, 2017. 20(3): p. 833–858.
81. Marinakis, V., H. Doukas, C. Karakosta, and J. Psarras. An integrated system for buildings' energy-efficient automation: Application in the tertiary sector. *Applied Energy*, 2013. 101: p. 6–14.
82. Sacks, R., M. Radosavljevic, and R. Barak. Requirements for building information modeling based lean production management systems for construction. *Automation in Construction*, 2010. 19(5): p. 641–655.
83. Babič, N.Č., P. Podbreznik, and D. Rebolj. Integrating resource production and construction using BIM. *Automation in Construction*, 2010. 19(5): p. 539–543.
84. Blazquez Cano, M., P. Perry, R. Ashman, and K. Waite. The influence of image interactivity upon user engagement when using mobile touch screens. *Computers in Human Behavior*, 2017. 77: p. 406–412.
85. Dehghani, M., K.J. Kim, and R.M. Dangelico. Will smart watches last? Factors contributing to intention to keep using smart wearable technology. *Telematics and Informatics*, 2018. 35(2): p. 480–490.
86. Barrios, D., R. Vergaz, J.M. Sánchez-Pena, B. García-Cámara, C.G. Granqvist, and G.A. Niklasson. Simulation of the thickness dependence of the optical properties of suspended particle devices. *Solar Energy Materials and Solar Cells*, 2015. 143: p. 613–622.
87. Ghosh, A., B. Norton, and A. Duffy. Daylighting performance and glare calculation of a suspended particle device switchable glazing. *Solar Energy*, 2016. 132: p. 114–128.
88. Mun, S., H.-U. Ko, L. Zhai, S.-K. Min, H.-C. Kim, and J. Kim. Enhanced electromechanical behavior of cellulose film by zinc oxide nanocoating and its vibration energy harvesting. *Acta Materialia*, 2016. 114: p. 1–6.
89. Atak, G., and Ö.D. Coşkun. $LiNbO_3$ thin films for all-solid-state electrochromic devices. *Optical Materials*, 2018. 82: p. 160–167.

90. Sun, J., X. Pu, C. Jiang, C. Du, M. Liu, Y. Zhang, Z. Liu, J. Zhai, W. Hu, and Z.L. Wang. Self-powered electrochromic devices with tunable infrared intensity. *Science Bulletin*, 2018. 63(12): p. 795–801.

91. Dulgerbaki, C., A.I. Komur, N. Nohut Maslakci, F. Kuralay, and A. Uygun Oksuz. Synergistic tungsten oxide/organic framework hybrid nanofibers for electrochromic device application. *Optical Materials*, 2017. 70: p. 171–179.

92. Dulgerbaki, C., N. Nohut Maslakci, A. Ihsan Komur, and A. Uygun Oksuz. Electrochromic strategy for tungsten oxide/polypyrrole hybrid nanofiber materials. *European Polymer Journal*, 2018.

93. White, T.J., L.V. Natarajan, V.P. Tondiglia, P.F. Lloyd, T.J. Bunning, and C.A. Guymon. Holographic polymer dispersed liquid crystals (HPDLCs) containing triallyl isocyanurate monomer. *Polymer*, 2007. 48(20): p. 5979–5987.

94. Bunning, T.J., L.V. Natarajan, V.P.T. Tondiglia, and R.L. Sutherland. Holographic polymer-dispersed liquid crystals (H-PDLCs). *Annual Review of Materials Science*, 2000. 30(1): p. 83–115.

95. Rauh, R.D. Electrochromic windows: An overview. *Electrochimica Acta*, 1999. 44(18): p. 3165–3176.

96. Lampert, C.M. Large-area smart glass and integrated photovoltaics. *Solar Energy Materials and Solar Cells*, 2003. 76(4): p. 489–499.

97. Piccolo, A., and F. Simone. Performance requirements for electrochromic smart window. *Journal of Building Engineering*, 2015. 3: p. 94–103.

98. Bouteiller, L., and P.L. Barny. Polymer-dispersed liquid crystals: Preparation, operation and application. *Liquid Crystals*, 1996. 21(2): p. 157–174.

99. Wang, K., J. Zheng, Y. Liu, H. Gao, and S. Zhuang. Electrically tunable two-dimensional holographic polymer-dispersed liquid crystal grating with variable period. *Optics Communications*, 2017. 392: p. 128–134.

100. Su, Y.C., C.C. Chu, W.T. Chang, and V.K.S. Hsiao. Characterization of optically switchable holographic polymer-dispersed liquid crystal transmission gratings. *Optical Materials*, 2011. 34(1): p. 251–255.

101. Hosseinzadeh Khaligh, H., K. Liew, Y. Han, N.M. Abukhdeir, and I.A. Goldthorpe. Silver nanowire transparent electrodes for liquid crystal-based smart windows. *Solar Energy Materials and Solar Cells*, 2015. 132: p. 337–341.

102. Kiruthika, S., and G.U. Kulkarni. Energy efficient hydrogel based smart windows with low cost transparent conducting electrodes. *Solar Energy Materials and Solar Cells*, 2017. 163: p. 231–236.

103. Dalapati, G.K., A.K. Kushwaha, M. Sharma, V. Suresh, S. Shannigrahi, S. Zhuk, and S. Masudy-Panah. Transparent heat regulating (THR) materials and coatings for energy saving window applications: Impact of materials design, micro-structural, and interface quality on the THR performance. *Progress in Materials Science*, 2018. 95: p. 42–131.

104. Mohamed, A.S.Y. Smart materials innovative technologies in architecture: Towards innovative design paradigm. *Energy Procedia*, 2017. 115: p. 139–154.

105. Marchwiński, J. Architectural evaluation of switchable glazing technologies as sun protection measure. *Energy Procedia*, 2014. 57: p. 1677–1686.

106. Xie, Z., X. Jin, G. Chen, J. Xu, D. Chen, and G. Shen. Integrated smart electrochromic windows for energy saving and storage applications. *Chemical Communications*, 2014. 50(5): p. 608–610.

107. Casini, M. Active dynamic windows for buildings: A review. *Renewable Energy*, 2018. 119: p. 923–934.

108. Casini, M. *9 - Dynamic glazing*, in *Smart Buildings*, M. Casini, Editor. 2016, Woodhead Publishing. p. 305–325.

109. Li, H., L. McRae, C.J. Firby, M. Al-Hussein, and A.Y. Elezzabi. Nanohybridization of molybdenum oxide with tungsten molybdenum oxide nanowires for solution-processed fully reversible switching of energy storing smart windows. *Nano Energy*, 2018. 47: p. 130–139.

110. Cai, G., X. Cheng, M. Layani, A.W.M. Tan, S. Li, A.L.-S. Eh, D. Gao, S. Magdassi, and P.S. Lee. Direct inkjet inkjet-patterning of energy efficient flexible electrochromics. *Nano Energy*, 2018. 49: p. 147–154.

CHAPTER 2

Perspective on Materials, Design, and Manufacturing of Electrochromic Devices

2.1 ELECTROCHROMIC WINDOWS

Energy efficiency in an architectural design has often been ignored as an opportunity for CO_2 reduction. However, recent developments in Europe have highlighted its importance. There are many "green" technologies that can be employed to create better buildings, which can lead to substantial economic savings. Energy-efficient glazing stands out as one of the most interesting possibilities, as it can regulate energy flows to enter or exit a building [1–3]. This means that external cooling or heating conditions must be exploited to create a comfortable indoor environment. Indoor energy efficiency can be reached with optimized glazing that allows regulated transmittance of solar energy and visible light. This type of glazing is often referred to as "smart" or "intelligent" and is based on "chromogenic" materials [4,5].

Electrochromism refers to the reversible change in optical properties when a material is subjected to electrochemical oxidation or reduction. Electrochromism has had a long history of fundamental and practical interest among architects [6]. Whenever a material showed significant color change it was designated as electrochromic (EC). EC glazing was proposed in 1984 and was introduced on the market during the 2010s. Since then, the number of EC installations and their current real-life uses have increased [7,8]. However, the recent interest in EC devices for multi-spectral energy modulation by reflectance and absorbance has extended the working definition of the materials used in these devices [9]. EC materials and devices are characterized based on their ability to transmit and reflect luminous (visible) and solar radiation. Currently, EC devices are being researched for change in color in the near-infrared, thermal-infrared, and microwave regions. The term "color" refers to the response of the detectors to these wavelengths and not just that of the human eye. From a commercial standpoint the interest in EC materials has been centered on three main types of product: windows, displays, and mirrors [10,11]. In general, EC devices have a two-terminal electrochemical cell configuration and employ at least one optically transparent electrode. In the earlier days of evolution, the attention to EC materials was directed to information displays. This application required high-contrast visible color changes against a diffusively scattered white background, which was incorporated into the electrolyte. EC materials for displays are required to exhibit multiple switching cycles. EC materials with high photopic coloration efficiencies (HPCE) will need lower charge densities that eventually affect the switching cycle. HPCE can improve the switching rate and enhance the battery life [12,13]. EC devices operate in the double-pass rather than transmission mode, which reduces the charge requirement by half to produce a given absorbance change. EC rearview mirrors have similar benefits. EC windows promise actively controlled and continuously tunable light transmission. Computer simulations of energy flows in buildings equipped with EC windows show significant energy savings in some locations and seasons, providing return on investment in a very short period [14–16]. In addition, factors related to comfort, privacy, glare, and fading have driven interest in EC window development for offices, homes, and automobiles [17–19].

This chapter presents an overview of EC windows and their unique materials, stability, and manufacturing requirements. This chapter also presents a discussion of device structures and provides several recent examples and multi-referenced compilations of EC material properties and current commercial development activities.

2.2 ELECTROCHROMIC WINDOW CONFIGURATIONS

There are presently three major EC window configurations, as shown in Figure 2.1. These configurations are solution phase, battery-like, and hybrid structures. In all the configurations, there are optically transparent conducting front and back electrodes for delivering current [3,20]. Glass is most commonly used as substrate material, but flexible foils of plastics (polyethylene terephthalate (PET)) are also alternatives. These substrates are coated with thin films of indium tin oxide (ITO). The battery-like configuration comprises a thin film of EC material and counter electrode materials which coat the optically transparent substrates.

The electrolyte, which is ionically conducting but electronically insulating, separates the electrodes. The electrolyte can be made up of either a polymer (gels) or a thin film. Polymer electrolytes lead to laminated sandwich EC structures while thin film electrolytes are the basis of all-thin film EC coatings. The thin film coatings are manufactured by deposition processes common to the architectural glass coating industry, e.g., sputtering, evaporation, and chemical vapor deposition (CVD). The battery-like EC windows have extended open circuit memory where continuous current is

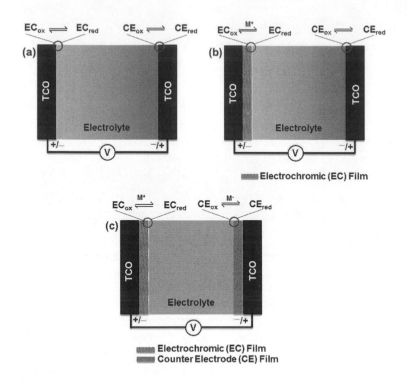

FIGURE 2.1 Schematic illustrations of electrochromic window configurations: (a) solution, (b) hybrid, and (c) battery-like. (Adapted with modifications from Rauh, R.D. *Electrochimica Acta*, 1999. 44(18): p. 3165–3176. [3]; Rauh, R.D., and S.F. Cogan. *Journal of the Electrochemical Society*, 1993. 140(2): p. 378–386. [23])

supplied to maintain the device in the colored state. Here the depth of coloration is proportional to the current density. Solution and hybrid EC windows fall into the self-erasing category, in which one or both of the color-changing EC materials is dissolved in a liquid or gel electrolyte, where it can easily diffuse. These approaches have been implemented in commercially successful EC mirrors [21,22].

The stoichiometry of EC window operation, like all devices, requires charge balance between the two electrodes such that charge inserted during the cathodic half reaction gets balanced by charge removed from the anodic side of the device. The generic reaction in EC windows can be written as [23]:

$$EC^n + yCE^m \leftrightarrow EC^{n-a} + yCE^{m+(a/x)} \qquad (2.1)$$

where EC and CE are the primary electrochromic and counter electrodes, respectively. In complementary EC windows, CE is an anode that is the coloring entity, which takes care of the overall coloration process. Here, n and m represent the oxidation states of the EC moieties in the bleached state, while x indicates the fact that EC and CE can be present in any stoichiometric ratio. The left side of the reaction indicates the bleached state, while the right side shows the colored state of the EC material. As shown in Figure 2.1, EC and CE can exist together in solution, or one or both may be immobilized as a thin film on the transparent conducting electrodes. Thin film electrodes need intercalation and de-intercalation of charge-compensating ions during the redox cycling.

Compared to displays and automotive rearview mirrors, the requirements of materials for EC windows are quite stringent [11,24]. As these devices function in the transmission mode, the combined EC and CE coloration efficiency should exhibit values that are twice that of displays or mirrors so that the same absorbance change per unit charge can be achieved. Such a stringent requirement underlines the importance of the CE reaction in EC windows in providing high bleached state transmittance and enhanced specific reversible capacity, preferably with complementary electrochromism [25–28]. Furthermore, the impact of environmental stresses on EC windows for architectural or automotive applications is expected to be greater than that of displays and interior mirrors. If the area of windows is large, then the switching capability becomes limited by the sheet resistance of the transparent electrodes that are used to deliver current. Finally, large area manufacturing processes are required so that architectural window products can be produced at lower cost per area than displays or even mirrors. Some of these requirements are relaxed for small area products like optical filters and eye wear [21].

2.3 MATERIALS FOR ELECTROCHROMIC WINDOWS

Materials for EC windows must provide an optical response that enables high transmittance in the bleached state and high diminution in the colored state. The transition between these states must be reversible according to Equation 2.1 over a well-defined redox region. For EC materials that show Beer's law behavior can be defined by a coloration efficiency $\eta\,(\lambda)$. The optical response of a complementary EC window is given by Equation 2.2 [3,23]:

$$A(\lambda) = A_{bl}(\lambda) + Q_T(\eta_1 - \eta_2) \tag{2.2}$$

$$A_{bl}(\lambda) = \alpha_1 d_1 + \alpha_2 d_2 + \eta_1 q_1' + \eta_2 q_2' \tag{2.3}$$

The absorption coefficients (α, cm^{-1}) and thicknesses (d, cm) of EC and CE in the bleached state determine the bleached state serial absorbance (A_{bl}) of the window, which can be expressed as a broadband optical density or at a specific wavelength. Q_T represents the transferrable charge within the stability region of EC and CE. The combined coloration efficiencies indicate the increase in absorbance over the bleached state during coloration process. For the sake of convenience, cathodic and anodic coloration are denoted by positive and negative coloration efficiencies, respectively. The bleached state of the window may also have residual coloration due to incomplete redox reactions of EC and CE which are denoted by residual charges q_1' and q_2'. This is considered a manufacturing defect [23].

Equations 2.2 and 2.3 can be used to quantitatively compare materials for EC windows. It is important to determine the absorption coefficient of EC materials in their bleached state as it indicates both the bleached state transmittance and the amount of charge available for coloration. For solution phase EC windows, the previously noted Equations can be expressed in terms of molar extinction coefficients of the bleached components.

2.3.1 Metal Oxide Based EC Windows

Application of synthetic chemistry greatly expands the selection of working electrode, electrolyte, and counter electrode materials for designing windows with desirable properties such as high coloration efficiency and rapid switching. The WO_3 electrode is a common EC layer used in smart windows. Amorphous thin films (a-WO_3) are generally prepared by evaporation and sputtering processes, as well as by chemical vapor deposition (CVD), electrodeposition, and sol-gel processes. However, they dissolve in acid media on long-term cycling or storage. This dissolution or chemical stability is morphology dependent. The stability can be enhanced in protic environments by shielding it with semi-porous hydrated Ta_2O_5 or copolymers of hydroxyethylmethacrylate (HEM) and 2-acrylimido-2-methyl-1-propanesulfonic acid (AMPS) [29,30]. As WO_3 is unstable in basic media, counter electrodes for proton based EC windows must also exhibit reversibility under acidic conditions of protic polymer and gel electrolytes (e.g., hydrated polyAMPS). The principal inorganic material known to satisfy this requirement and possessing apt complementary optical properties is Iridium oxide (IrO_2) [3,31,32]. The stoichiometric reaction describing the operation of the WO_3-IrO_2 EC window can be given as [3]:

$$WO_3 + yH_{x/y}IrO_2 = H_xWO_3 + yIrO_2 \qquad (2.4)$$

The left side of the reaction indicates the bleached state while the right side indicates the colored state of the EC window.

Other oxides, such as Nb_2O_5 or crystalline WO_3 [3,33,34] that are stable to acid solutions, provide weak cathodic electrochromism (this, again, dependent on particle morphology) and can be used as counter electrodes in non-complimentary windows [35,36]. However, in the current stage of development they are unsuitable for producing windows with a large visible dynamic range.

EC windows employing WO_3-IrO_2 have been extensively studied. In these studies, both hydrated polymers (mostly polyAMPS) and thin protonic films (mostly Ta_2O_5.nH_2O) have been evaluated as electrolytes [23,37,38]. Hydrated stable polyAMPS proton conductors were prepared by casting using an aqueous solution at a given humidity. Bubble defect free, low scattering EC window laminates were fabricated by employing modified polyAMPS formulations in autoclave and photo-polymerization processes. Windows which were fabricated using low density [3,39,40] (deposited by electron beam evaporation and sputtering method) with intrinsic or intentionally added water vapor showed $>10^5$ room-temperature color and bleach cycles. Here, IrO_2 has a small residual absorption in the reduced state and the dynamic range will be a function of IrO_2 capacity. Thus, thicker electrodes of IrO_2 are required to provide the necessary charge density (Q_T) to color to the very high absorbance values with lower bleached state transmittance.

Another oxide that has gained attention as an aqueous counter electrode is nickel oxide, which is stable in basic solutions but dissolves in acid. The EC mechanism here is anodic due to the absorbance of Ni^{3+} centers [41,42]. Window set-ups having WO_3 as a working electrode or EC layer, as electrolyte, and $Ni(OH)_2$ have been reported with $>10^4$ color to bleach reversible cycles at 85°C [43]. The solid electrolyte $Ta_2O_5 \, nH_2O$, although acidic, prevents NiO dissolution [44]. The water in the thin film layers can be used as a "resource" to color the device on the first cycle and can have a role in providing the reserve of fresh electrochemically exchangeable charges on extended cycling. These devices require constant hydration to maintain proton conductivity, as the key water loss mechanism is electrolysis. This is because the Ni^{2+} oxidation state occurs close to the O_2 evolution voltage. The electrolyzed water can cause WO_3 dissolution, and thus there has been considerable research interest in aprotic WO_3-based EC windows [45]. WO_3 has a unique property among oxide insertion compounds of forming stable hydrogen and alkali bronzes. Li^+ is a widely researched candidate (as cation) for counter electrodes in EC devices due to its excellent mobility in WO_3. The availability of Li^+ solid polymer electrolytes used in Li-ion battery research has provided several new solvent chemistries that are mechanically stable and have ionic conductivities $>10^{-3}$ S cm^{-1} at room temperature. For counter electrodes, Li insertion compounds of $LiCoO_2$, $LiCrO_{2+x}$, NiO, V_2O_5, and a wide range of mixed metal oxides that are complementary to WO_3 in their overall coloration [3,45] are available. An ideal counter electrode should exhibit coloration efficiency measurements of about 25 mC cm^{-2} to color a-WO$_3$ to an absorbance of 1.0 at 600 nm. Counter electrode materials can be classified by their reduced state absorption coefficients, which may range from about 10^5 cm^{-1} for a direct band gap semiconductor to $<10^3$ cm^{-1} for a high band gap semiconductor in the broadband mid-visible spectrum [3,46].

Table 2.1 shows the specific capacities of various oxides taken from the battery literature [46]. The thickness needed for storing the required charge for WO$_3$ or other EC material can be calculated from these data. Some of these materials have specific capacities >0.5 mC cm^{-2} nm^{-1}, which means only layers on the order of 50 nm would be sufficient. Such a counter electrode material in its reduced form can have an absorption coefficient of 2×10^4 and can add only 0.1 absorbance unit to the bleached state device.

TABLE 2.1   Electrochemical Characteristics of Li Insertion Oxides for Counter Electrodes

No.	Counter Electrodes	Mid Voltage (V)	Reversible Domain x (Li)	Density (g cm^{-3})	Specific Capacity μCcm^{-2} nm^{-1}
1	LiV$_3$O$_8$	2.8	0–2.5	~3.2	0.27
2	LiCoO$_2$	4.0	0.3–1.0	4.2	0.21
3	LiNiO$_2$	4.0	0.6–1.0	3.6	0.14
4	LiMn$_2$O$_4$	4.1	0–1.0	4.2	0.22
5	LiCrO$_2$	3.7	0.8–1.0	3.9	0.31
6	LiVO$_2$	3.0	0.7–1.0	3.7	0.12
7	LiCr$_3$O$_8$	2.2	0–4.0	4.2	0.79
8	WO$_3$	2.3	0–0.57	7.16	0.21
9	TiO$_2$	1.3	0–0.7	4.95	0.42
10	MoO$_3$	2.5	0–1.5	4.69	0.47
11	Cr$_3$O$_8$	2.8	0–1.5	~5.0	0.25
12	V$_2$O$_5$	3.2	0–1.0	3.35	0.18
13	V$_6$O$_{13}$	2.8	0–8.0	3.9	0.59
14	MnO$_2$	3.0	0–1	5.03	0.56

Source: Patil, R.A. et al. *Solar Energy Materials and Solar Cells*, 2016. 147: p. 240–245. [18]

Sputtered thin coatings of a chromium oxide (CrO_2) are lithiated by a "dry" evaporation process and heat treated to form a thin film with composition of "Li_zLiCrO_{2+x}," where z denotes electrochemically transferable Li. While the stoichiometry of these films is not precisely known, they have suitable properties for a counter electrode, which include net anodic coloration and a reversible charge storage capacity of >0.5 mC cm^{-2} nm^{-1}. Thin film windows are fabricated using an evaporated ion conductor with approximate stoichiometry $LiBO_2$ [47]. Over 10^5 color to bleach reversible cycles have been demonstrated at 60°C without significant change in the colored and bleached spectra. The windows are sensitive to moisture, need to be protected by an outer airtight layer such as SiOC:H, or should be operated in an inert atmosphere environment. Chemically adsorbed water can be present in these films despite a processing temperature of 175°C. Coloring counter electrodes include hexacyanometallates, such as Prussian blue (PB) and oxidatively doped conductive polymers, such as polyaniline (and their composites with PB). The coloring electrode also includes extensive classes of redox-active dyes that are reducible in protic environments to their "leuco" forms. It should be noted that it is more difficult to attain uniformity and long term durability in organic films than in inorganic materials that are obtained by physical or chemical evaporation processes. Leventis and Chung in their studies have described methods for surface confinement of dyes through the incorporation of vinyl or $-OSi(Me)_3$ groups and subsequent polymerization [48].

Hybrid self-erasing EC windows were first commercially produced by Asahi Glass Co., Ltd. in Japan. A typical structure of a hybrid self-erasing EC window consists of WO_3 deposited on ITO as the EC electrode [49]. LiI was used as the counter electrode that was dissolved in a solution of l-butyrolactone and a polymer thickening agent. The WO_3 electrode produces Li_xWO_3 on application of current and I_3^- in solution. The compounds reversibly react when the current is removed. EC windows with crystalline WO_3 can provide 60%–90% near-infrared reflectance modulation on reduction. It was seen that the formed Li_xWO_3 can exhibit variable electron density and a plasma edge that can be moved throughout the solar spectrum. This EC window structure had $LiCoO_2$ counter electrodes and $LiNbO_3$ electrolytes. The optical modulation was observed to be shifted towards the near-infrared for c-WO_3, and the charge density required to achieve similar degrees of optical modulation in the visible space was considerably higher in c-WO_3 than a-WO_3. This necessitated the requirement for thicker counter electrodes (in case of c-WO_3) for daylight attenuation applications. As a result, there is a greater burden on the counter electrodes for providing optical transparency in its bleached (reduced) state. Studies have shown that to attain full range of infrared reflectance at least $Li_{0.5}WO_3$ stoichiometry needs to be achieved while employing c-WO_3. The EC device structures, which require infrared passive counter electrode materials like a-WO_3, employ current collectors at micro-grid level rather than ITO or any other transparent oxide films that are infrared reflective [50–53].

2.3.2 Optical Absorption in EC Oxides

The mechanism underlying the optical absorption in EC oxides has been investigated for several years. The origin of optical absorption in EC oxides is a complex mechanism involving various factors and only a simplified view, capturing the most salient features, is explained here. The complications arise from the fact that there is a lack of understanding of the crystal structure and processes involving oxygen deficiency and inclusion of mobile ions and water molecules. Nevertheless, a detailed explanation has been presented for WO_3 structures, while much less is known for NiO and other EC oxides [54,55]. Insertion and extraction of protons (h+) and electrons (e^-) in WO_3 can be described by the following electrochemical reaction [56,57]:

$$[WO_3 + h^+ + e^-] \leftrightarrow [HWO_3] \qquad (2.5)$$

Here, h+ can be replaced by Li^+ or some other ion. Full reversibility can be achieved for certain partial reactions, which implies that the colored material can be written as $h_x WO_3$ or $Li_x WO_3$ with $x < 0.5$ [58,59]. For NiO, the corresponding electrochemical reaction can be expressed as:

$$[Ni(OH)_2] \leftrightarrow (NiOOH + h^+ + e^-) \qquad (2.6)$$

The previous reaction is believed to take place on hydrous grain boundaries. It is confined to the surface when Li^+ is the mobile ionic species. For reactions expressed in Equations (2.5) and (2.6), the left and right side of the reaction indicates the bleached and the colored state, respectively [60,61].

The fact that most EC oxides are assumed to be constructed from octahedra-like structural units makes it feasible to put forward a schematic model for the occurrence of cathodic and anodic electrochromism. The oxide structures can be characterized by O2p bands well separated from the metal d band. The octahedral symmetry leads to a splitting of the latter band into sub-bands. This has been designated as e_g and t_{2g} in Figure 2.2 where three cases of relevance for EC oxides are shown. Specifically, the left section in Figure 2.2 for $H_x WO_3$ shows an O2p band that is separated from a split d band by an energy gap. For stoichiometric WO_3 the O2p band is full and the d band is empty, and the band gap is wide to emit luminous transmittance for thin films. The insertion of ions and charge balancing by electrons partially fills the d band. Intercalation of ions and charge balancing electrons results in partial filling of the d band along with optical absorption. The middle panel of Figure 2.2 relates to anodic coloring EC oxides, such as IrO_2 and over stoichiometric NiO_x (with $x > 1$), which can be described as having unoccupied t_{2g} states. The ions and electrons fill these states to the top of the band indicating that the oxide displays a gap between the e_g and t_{2g} sub-bands. The oxide becomes transparent, assuming that the band gap is sufficiently large. Finally, the right section of Figure 2.2 shows $V_2 O_5$ exhibiting both cathodic and anodic characteristics and having different electronic structure. Here the d band exhibits a narrow split-off part in the band gap, exiting from the octahedral coordination.

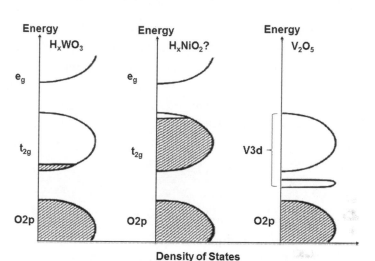

FIGURE 2.2 Schematic presentation of band structures for different types of EC oxides. Shaded regions indicate filled states, and E is energy. The relevant chemical species are indicated at the top, while some uncertainty prevails, indicated as (?), for the Ni-containing species. (Adapted with modifications from Granqvist, C.G. et al. *Electrochimica Acta*, 2018. 259: p. 1170–1182. [21])

The insertion of ions and electrons into the V_2O_5 structure can fill this narrow band so that the optical band gap is widened, which results in electrochromism in V_2O_5 as well as band gap broadening upon photo-injection of H_2 [21,62–64].

In the case of WO_3, the insertion of electrons becomes localized on W sites, and some of the $W6^+$ sites turn into $W5^+$ sites. The inserted electrons can get sufficient energy to be shifted to a neighboring site by the absorption of a photon. Transfer between these two sites can be denoted by i and j, which can be described as [65,66]:

$$W_i^{5+} + W_j^{6+} + \text{photon} \rightarrow W_j^{6+} + W_i^{5+} \tag{2.7}$$

Here, the electrons are believed to enter localized states positioned between 0.1 and 0.2 eV below the conduction band. The atoms are dislocated so that a potential well is formed and pronounced electron-phonon interaction leads to the formation of polarons with an extent of 0.5–0.6 nm. The effect of W^{4+} has been emphasized in some published literature. Polaron-induced optical absorption in WO_3 produces a wide band centered at $\lambda = 0.85$ μm. This extends into the luminous region of the spectrum so that thin film achieves a transmission of blue color. A detail theory on the spectral dependence of the absorption is described in references 67–69.

2.3.3 Mixed Oxides

Mixed oxides have shown to have better EC properties than pure oxides. One of the reasons for these enhanced properties has been the additional optical transitions that are made possible and that can produce an intermixing of polaron-induced absorption bands. Many studies have been reported recently on oxides based on W-Ti, W-V, W-Ni, W-Mo, W-Nb, W-Ta, Mo-V, Mo-Ti, Mo-Nb, Mo-Ce, Ti-Zr, Ti-V, Ir-Ta, Ir-Sn, Ni-Al, Ni-Ti, Ni-V, etc. Blending between WO_3 and an organic substance is another option that is being explored. Recent studies on W-TiO_2 and Ni-IrO_2 blends have shown enhanced electrochemical durability. Further enhancement in EC properties can be accomplished using ternary oxide compositions. This was demonstrated by using thin films of $W_{1-x-y}Ti_xMo_yO_3$ with x < 0.2 and y < 0.2. The study showed that when Ti was combined with Mo, electrochemical durability along with color neutrality could be achieved with WO_3 [70–72].

However, electrochemical durability was affected with the addition of excess Mo, and it was appropriate to maintain the Mo composition below 6 atomic weight percentage (at.wt.%). Complex anodic coloring EC oxides have been investigated, and interesting results have been reported for oxides of Ni-Al-Li, Ni-W-Li, Ni-Fe-C, Ni-Zr-Li and Ni-Ti-Li, as well as on Ni-LiPON [73–77].

2.3.4 Organic and Organometallic Electrochromic Windows

Recent progress in organic electrochromism has led to interesting window systems.

For instance, Leventis and Chung provided data for cathodic and anodic coloring layers comprising a "surface confined" viologen polymer and a poly-pyrrole-PB composite, respectively. Both these layers were separated by a K_2SO_4, poly(vinylpyrrolidone) gel electrolyte. Coloration efficiencies at 600–650 nm wave length were observed as 360 cm^2 C^{-1} and −129 cm^2 C^{-1} for the cathode and anode, respectively as compared to 45 cm^2 C^{-1} for the H_xWO_3-IrO_2 devices at 633 nm wavelength [38,48].

Poly(3,4-ethylenedioxythio-phene) (PEDOT) is a conducting polymer that has emerged as a robust cathodic coloring component. The oxidized form of PEDOT absorbs in the near-infrared region and appears colorless while the reduced form absorbs in the mid-visible region. Reynolds and team have reported a variety of N-substituted bis-EDOTcarbazoles that can be electrochemically polymerized to form highly anodic coloring films that are colorless in the reduced state [78]. These window structures employing a Li$^+$ gel electrolyte showed a remarkable coloration efficiency of 1,409 cm^2 C^{-1} at 650 nm. Sub-second switching rates were achieved due to the low charge density requirements (<1 mC cm^{-2} at 650 nm wavelength), along with 10^4 full cycles with 60% of the dynamic range retained. An organic-based microwave window was demonstrated, where the front electrode (polyaniline/camphor sulfonic acid) emits microwaves in reduced state and mitigates in oxidized conductive state. The counter electrode was kept as $Li_xMn_2O_4$, which is passive to microwaves. Electrodes were made from Cu/Au grids on X-band transmitting RT-Duroid polymer. Polyacrylonitrile Li$^+$ conducting polymer was used as the electrolyte. A maximum modulation of 4.8 ± 42% at 10 GHz was observed. To achieve high microwave attenuation, thicker films (about 38 μm) of polyaniline were required. This thickness range was easier to attain using a solution cast for organic materials than for vapor deposited inorganic films like c-WO$_3$.

Solution-phase EC windows are organic systems with co-dissolved redox pairs. The typical mixture contains N,N,N′,N′ tetramethyl-p-phenylene diamine (NTMPD) and

N,N'-di-n-heptylviologen (NHV) dissolved together in an organic electrolyte. Passing an electrical current causes oxidation of NTMPD at the anode and reduction of NHV at the cathode, both of which are blue in color and therefore complementary. Withdrawal of the current causes NHV+ and NTMPD+ to recombine to the colorless starting materials. The transmittance of this EC structure at 633 nm was observed to be <80% at 1.2 V. The maintenance currents approached 10 mA cm^{-2}, which could be controlled by increasing viscosity, but at the expense of the switching rate. Solution-based windows are claimed to be less expensive compare to battery-like designs. Currently, prototypes of 15 inch^2 are being manufactured that are undergoing extensive environmental testing by Gentex Corporation [79,80].

2.4 PERFORMANCE CONSIDERATIONS

In most applications, EC windows are continuously exposed to solar radiation, which results in thermal cycling issues arising from interior-exterior thermal gradients. While there have been several studies on thermal durability of EC window prototypes, accelerated standard thermal cycling tests have been designed at independent test facilities like National Renewable Energy Laboratory (NREL). The instabilities in EC windows can result from four different sources: electrochemical, photochemical, chemical, and mechanical. Electrochemical instabilities arise when EC materials go beyond potential voltage of stability where they are irreversibly oxidized or reduced. Photochemical instability results from electronic transition of the window constituents due to light absorption, which is followed by relaxation of the excited states into irreversible product-forming channels. Prolonged photochemical stability of colored organic compounds in direct sunlight is difficult to achieve. Photoelectrochemical compounds made up of metal oxides and oxidizable organic compounds or H$_2$O can react with each other and lead to product formation at oxide electrode-electrolyte interfaces. Corrosion can occur due to irreversible chemical reactions between adjoining dissimilar window components resulting in chemical instability. Both electrochemical and chemical instabilities are expected to increase with temperature [81].

When a voltage is applied to the optically transparent electrodes, the voltage distribution at the electrode-electrolyte interfaces is not uniform. It is possible that all voltage can be dropped across just one of the interfaces as a result. The voltage drop is maximum at the edge contacts in large area windows and is then spread toward the center.

Over-polarization can result in irreversible reaction products of electrode and electrolyte constituents. For example, high levels of Li insertion or extraction in metal oxides result in irreversible phase changes. Reduction of WO_3 with $Li_{0.7}WO_3$ can lead to irreversible non-stoichiometric WO_3 and insert Li_2WO_4 [82]. Ceder and Ayindol have demonstrated that Li insertion into $LiMO_2$ compounds (M = Mn, Fe, Co, and Ni) makes it thermodynamically unstable against M reduction [83]. This could lead to gradual capacity fade. Excessive positive polarization of open, high Li mobility lattice structures like $LiCoO_2$ can cause Li extraction to a point of irreversible crystal structure damage [46].

To counter these problems, it is preferable that EC material and counter electrodes be present in stoichiometric excess to extend bleached state transmittance, so that the system is not stretched to a point of irreversibility and the polarization during switching is kept to a minimum. The required polarization is thermodynamically determined and can be predicted from cyclic voltammetry and coulometric titrations of the device assembly and from the oxidation states of the respective EC and counter electrode components. It is desirable to have polarization beyond these potential limits to increase the switching speed, but this also increases the probability of premature failure [23,38].

As an example, Figure 2.3 shows a theoretical schematic diagram of the electrochemical stability ranges relevant to the operation of a $Li_xWO_3|Li_2O-B_2O_3-Li_yCrO_{2+x}$ all thin film window system.

Irreversible regions caused by the over-reduction of WO_3 and by the over-redox-reactions of Li_yCrO_{2+x} are indicated in the

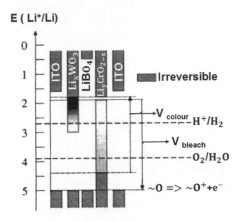

FIGURE 2.3 Schematic presentation showing the potential regions of stability for the $Li_xWO_3|Li_2O\text{-}B_2O_3\text{-}Li_yCrO_{2+x}$ window system. Irreversible regions of over-reduction of EC and counter electrode are shown along with degradative reduction of ITO. Over-oxidation of Li_yCrO_{2+x} is also presented. The potential region of stability for water is superimposed, and components which lie at potentials outside of that region are expected to undergo hydrolysis. Also shown are the maximum coloring and bleaching voltages, assuming a perfect stoichiometric match between EC and counter electrode and 100% transferrable charge. An oxidative decomposition of >5 V is indicated for the oxides except for Li_yCrO_{2+x}. (Adapted with modifications from Rauh, R.D. *Electrochimica Acta*, 1999. 44(18): p. 3165–3176. [3])

diagram. Transparent oxide electrodes undergo reduction in the presence of Li^+ at potentials of about -1.5 V. Removal of electrons from the valence bands of transparent oxides is expected to occur only at voltages higher than 5 V. A particular advantage of WO_3 is that it can act as a dielectric insulator when fully oxidized, blocking electron pathways on extreme polarization during bleaching. Other considerations are the effects of any moisture content on the thin films that may have been introduced during the fabrication or post-processing. In hermetically sealed $Li_xWO_3|Li_2O\text{-}B_2O_3\text{-}Li_yCrO_{2+x}$ test devices, it has been observed that switching speed varies with cycle life, which is reflected as a slowing performance in this system and can be attributed to the loss of water by electrochemical or corrosion pathways. The presence of water or moisture is difficult to remove from porous oxides [84–88].

Thermo gravimetric studies of $Ta_2O_5nH_2O$ indicate that n = 1.7 at 330°C. Studies have shown that water gets chemically adsorbed in a-WO_3 at >300°C [43]. The effect of moisture/water on the operation of Li-based thin film EC windows is still a researched topic and debatable. Dahn and team have shown operability of Li metal oxides with Li insertion potentials in concentrated aqueous solutions of Li salts. Recent studies have shown that nuclear reaction analysis (NRA) of a thin film Li-based device to track Li ion migration cannot conclusively rule out the presence of residual water. The maximum bleaching voltages and coloration for this device are shown in Figure 2.3. Here, it is assumed that neither EC material nor counter electrode is a limiting factor. However, if water is present, driving EC and counter electrode materials beyond the water decomposition limits can produce gases or reactions at the electrodes with residual water [43].

Devices with aqueous or organic electrolytes or organic EC layers are expected to be more sensitive to over-polarization due to their narrow decomposition limits of the constituents (e.g., irreversible "over-oxidation" of conductive polymers). The interaction between organic electrolytes and the surface of EC oxides and/or counter electrode layers also provides sites for decomposition similar to "capacity fading" observed in rechargeable Li-ion batteries. This mechanism affects the cycling of some thin film metal oxide insertion electrodes in organic electrolytes. However, these mechanisms are expected to be less critical in all inorganic, thin film EC windows. Ion conducting polymers and gels cannot be employed at temperatures beyond 60°C. Raman spectroscopy analyses have shown that systems like polyAMPS undergo hydrolytic decomposition at temperatures of 90°C. In addition, ITO gets etched at elevated temperatures by acidic media such as polyAMPS, resulting in a gradual loss in conductivity. For such acidic media, transparent oxides or metallic micro-grids are preferred [89,90].

2.5 MANUFACTURING CONSIDERATIONS

The actual optical modulation band achievable through reversible operation of a battery-like EC window is closely associated with the manufacturing process. Here, the concept of "transferrable" charge in the window after fabrication is important, as indicated in Equation 2.2.

In Li-ion battery-like configurations, the electroactive Li ion is usually introduced by dry, chemical, or electrochemical pre-reduction of EC material or counter electrodes during processing. Similarly, organic EC windows, either battery-like or solution-based, require assembly with the EC components in the proper oxidation state to ensure maximum available charge for transfer during operation. In complementary EC windows, residual absorption will occur if the counter electrode is incompletely reduced when EC materials get fully oxidized (see Equation 2.3). In aqueous-based EC windows, one of the electrodes gets electrochemically pre-reduced with H_2, or with a chemical reducing agent. Some EC windows with aqueous components have been reported that are charged after fabrication through the electrolysis of water in the electrolyte. This gives the benefit of completely oxidizing or reducing the limiting electrode and compensating the amount of charge in the non-limiting electrode. This mechanism is apparently seen for WO_3|Ir oxide and WO_3|Ni oxide systems and it could simplify the processing of these structures. A typical charging reaction on polarizing the hydrated, bleached device can be expressed as:

$$H_2O + (2/x)WO_3 + Ir^{+3}(OH)_3 \rightarrow (2/x)H_xWO_3 + Ir^{4+}(O)(OH)_3$$

$$(2.8)$$

Processing of all thin film systems requires careful process control to prevent short circuits from forming during production. Any intrinsic defects can lead to poor open circuit memory or to a total inability to cycle the structures. Large area (about 0.18 m²) all thin film WO_3|Ta_2O_5 nH_2O|H_xNiO_2 windows based on protic systems have been demonstrated recently. There are some studies that indicate that leakage current densities decrease with the area of windows, which can be attributed to the edge effects. Li ion thin film window structures of 25 cm^2 area have been produced with >80% yield. To compete in the architectural coatings markets, thin film EC glass must be scaled up by a high-rate in-line process. Line speeds of 500 cm min^{-1} are typical for coated glass. To achieve this speed with the required uniformity for all layers, processes like high rate sputtering would be required. For all thin film devices, suitable Li ion electrolytes and Li-loaded electrodes made by high rate dc magnetron sputtering are still not available. Value-added products, such as sun roofs and skylights, can offer higher profit margins with its capital/labor intensive and lower-rate batch processing. However, it must be noted that innovations in production are necessary to address the target markets. These are the key issues in scaling up EC materials for smart windows [3,91].

2.5.1 Mechanically Flexible EC Windows

Several designs have been proposed in the literature for making flexible thin film ECs. The present section provides information for one of these designs, specifically for mechanically flexible foils for EC windows that can be produced by low-cost high-throughput roll-to-roll processing. The design is schematically presented in Figure 2.4 and comprises (a) a 0.175-mm-thin PET foil with a transparent conducting In_2O_3:Sn film and a film of EC WO_3, (b) another PET foil with an In_2O_3:Sn film and a film of EC NiO, and (c) a polymer electrolyte connecting the two EC oxides. The foil can be employed in different ways, i.e., it can be either deployed onto the surface of an existing glass pane or it can be held between two glass panes in the glazing. This approach effectively alters it from a double-glazed to a triple-glazed structure without significantly increasing the areal density. It can also be used for glass lamination as delineated in the left-hand part of Figure 2.4. Some of these layered structures can give added functionalities to the EC device and impart shielding from impact, burglary protection, etc. [21].

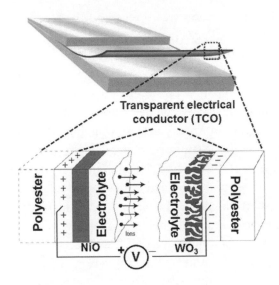

FIGURE 2.4 Principle design of a thin film-based EC device. Arrows indicate ion transfer when a voltage is applied between the transparent electrical conductors. The entire thin film can be employed to laminate glass panes, as shown in the left-hand part. (Adapted with modifications from Granqvist, C.G. et al. *Electrochimica Acta*, 2018. 259: p. 1170–1182. [21])

Figure 2.5 provides insights into the optical properties of an EC foil. Figure 2.5a demonstrates that the transmittance can be modulated within a wide range. Figures 2.5b and c show the optical modulation of the individual WO_3-based and NiO-based components after the device has been disassembled. It was observed that the oxide films are optically complementary with the WO_3-based film showing cathodic coloration mainly in the long wavelength part of the luminous spectrum. The NiO-based films exhibited anodic coloration mainly at lower wavelengths. EC foils for glazing are sensitive to UV irradiation, which can give a photochromic (PC) in addition to the EC effect. If so desired, PC effects can be avoided by coating the WO_3 with Ta_2O_5. Photochromism has been also observed in Ni-oxide based films. Color bleach dynamics is an important property for EC devices. For smaller area EC devices, the

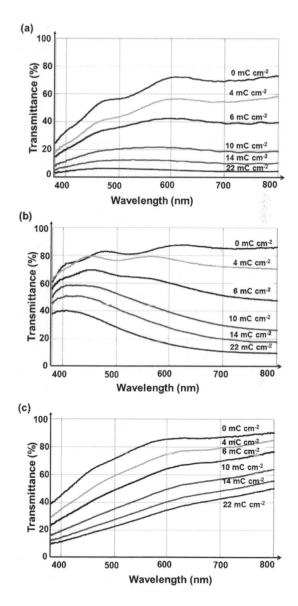

FIGURE 2.5   (a) Spectral transmittance for an as-prepared EC foil device of the kind shown in Figure 2.4, (b) data for the WO-based and (c) NiO-based parts of the device. (Adapted with modifications from Granqvist, C.G. et al. *Electrochimica Acta*, 2018. 259: p. 1170–1182. [21])

rate of change of opaque to transparent states, and vice versa, can be as little as a few seconds, but for large area devices, this rate is slower as the requirement of electrical charge that is inserted or extracted through electrical contacts ("bus bars") at one or, preferably, several of the device's edges is higher. The often unwanted feature that EC devices exhibit is called the "Iris effect" where the color change is faster at the edge than in the center. But this effect can be compensated via the electrical drive circuitry. The range of optical tuning depends on the kind of application of the EC foils. For instance, in buildings it is desirable to have glazing with large bleached-state transmittance and anti-reflection coatings. They should not display excessive light scattering (known as "haze"). The glare control is an important feature, and it is possible to decrease the colored-state transmittance through superimposed foils. Thus, if the transmittance is 10% in the opaque state for an EC foil, then two superimposed foils will give a transmittance of only about 1%. Besides optical properties, long-term durability is another essential parameter [21,92–95].

Thin film type EC devices include a layer of a polymer electrolyte, which can be functionalized by nanoparticles. EC devices use a "model electrolyte" of polyethyleneimine-lithium bis(trifluoromethanesulfonyl)imide (PEI-LiTFSI). In these types of electrolytes similar functionality with nanoparticles can be implemented. Another option is based on the employment of nanoparticles of transparent conducting oxides such as Sn doped In_2O_3 (ITO). In this scenario, it is possible to achieve near-IR plasmon-based absorption and diminish T_{sol} without significantly affecting T_{lum}. This is an important attribute for EC glazing that is used in warmer climates. Studies have proven that strong near-IR absorption occurs when the number of nanoparticles is increased. Using 7 wt% of Sn doped ITO, the EC foil has $T_{lum} = 83.3\%$ and $T_{sol} = 56.3\%$, while the electrolyte remains free from haze [96,97].

The measured data on $T(\lambda)$ can be obtained with quantitative calculations using detailed theoretical understanding of the properties of Sn doped ITO and by accurate descriptions ("effective medium models") for the optical properties with suspensions of nanoparticles [97]. The EC foil can be produced by a roll-to-roll web coating manufacturing technique, which is known for combining low cost with high through-put. This thin-film deposition process can be combined with continuous lamination of WO_3 coated and NiO coated PET foils by use of the polymer electrolyte. The end products are large flexible sheets that can be cut to any shape and/or size after which "bus bars" are applied. Consequently, manufacturing of the final foil device such as EC-based glazing can be detached from the site for foil production [97,98].

Figure 2.6 shows installations of large area glazing with EC-based foil laminated between glass panes. The photos of two of the windows in the upper panel appear to be opaque and distinctly different from the adjoining transparent window. The rate at which the windows transit from opaque to colored states is about 10 min, which is suitable for allowing the eye to light-adapt. No visible haze or other defects were observed. Figure 2.6b shows a building with thin film EC glazing on the lower two floors [99].

2.6 TOWARDS HIGH PERFORMANCE AND LONG LIFE EC GLAZING

Long-term durability or cycling stability is an important requirement for EC devices. This electrochemical cycling property has been widely investigated [100–104]. Durability or cycling stability is the ability to maintain charge transfer between the two thin film electrodes in an EC device for thousands of cycles without performance fading, resistance against degradation by solar irradiation on exposure for prolonged times, robustness against chemical attacks especially with regard at the interfaces, good shelf-life, etc. Recent work studied the decrease in the charge density Q- and the subsequent changes in the coloration for films of EC NiO and WO_3 [105–107].

Figure 2.7 provides the data recorded under cyclic voltammetry (CV) cycling for up to 10^4 cycles. Figure 2.7a shows current

FIGURE 2.6 Examples of (a) interior and (b) exterior views of EC glazing based on thin film deposition technology. (Adapted with modification from Granqvist, C.G. et al. *Electrochimica Acta*, 2018. 259: p. 1170–1182. [21])

density as the voltage applied to the films was swept between pre-set boundary points. The encircled region (area under the curve) corresponds to charge density exchange. The evolution of change in charge density is given in Figure 2.7b. It was observed to be an almost linear drop when expressed on logarithmic plots of Q versus x. Curve fitting analysis shows that a power-law or a stretched-exponential expression could accurately represent

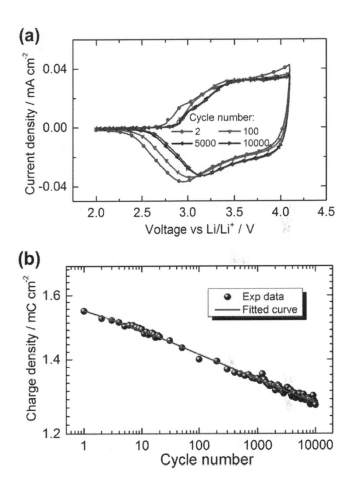

FIGURE 2.7 (a) Current density vs. voltage during long-time electrochemical cycling and (b) related charge density exchange for a NiO-based EC film immersed in a Li ion conducting electrolyte. Dots in Figure 2.7 (b) show experimental data and the fitted curve drawn to obtain Equation 2.9. (Adapted with modifications from Granqvist, C.G. et al. *Electrochimica Acta*, 2018. 259: p. 1170–1182. [21]; Wena, R.-T. et al. *Applied Physics Letters*, 2014. 105(16): p. 163502. [105])

the decline of the charge density. These curve fittings can be represented using the following expression for the power law:

$$Q = A_2 + \frac{A_1 - A_2}{1 + (x/x_0)^P} \qquad (2.9)$$

Here, A_1 and A_2 are the initial and final charge capacities, p is a kinetic exponent obtained from the curve fit, and x_0 is the cycle number at which the charge density Q has declined to an average of its initial and final value. The fitting parameters can be co-related to the film composition and voltage range for ion insertion/extraction. The underlying model for degradation can be corroborated with diffusive chemical kinetics, and it also involves a variety of diffusion-limited chemical reactions. However, the role of individual components affecting the degradation process is still poorly understood [106,108].

Recently it was reported that degraded EC films could be regenerated so that they could recover their original properties. This may open ways for new technologies to re-use EC devices for a long time. These findings can also help other relevant ionics-based devices such as electrical batteries. Most of the rejuvenation or regeneration studies have so far been carried out on WO_3 and have included galvanostatic and potentiostatic treatments. This rejuvenation or regeneration process can be performed multiple times and has been established for other EC electrode materials, such as TiO_2 and MoO_2 [109,110]. Figure 2.8 shows studies on sputter deposited WO_3 thin films immersed in a Li-conducting electrolyte and serves as a clear demonstration of rejuvenation. Figure 2.8a shows data on electrochemical cycling of a film after the initial and the 400th cycle in a given voltage interval. An apparent drop in the charge density exchange was seen. Figure 2.8c corresponds to the decrease of the optical modulation.

Device degradation can be attributed to the trapping of Li ions. This is based on the assumptions that WO_3, which is a network of connected sites with low inter-site barriers, allows fast ion diffusion throughout the host material and other sites with higher energy barriers that are capable of trapping diffusing ions. The trapping

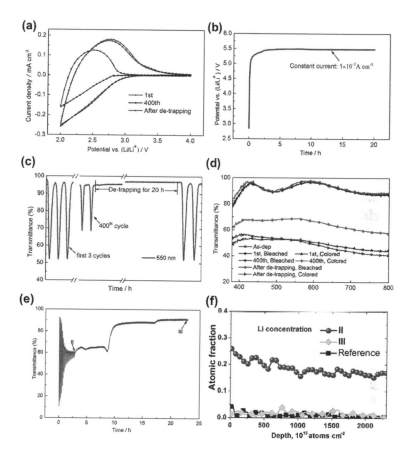

FIGURE 2.8 (a) Galvanostatic regeneration of an EC WO film, showing electrochemical cycling performed in a Li-conducting electrolyte, (b) voltage variation during galvanostatic expulsion of trapped ions, (c) optical transmittance in mid-visible region, (d) transmittance in the entire visible wavelength range and spectral transmittance overlap partly for bleached films, (e) Mid-visible optical transmittance vs. time during degradation and galvanostatic regeneration of WO film, and (f) ToF-ERDA depth profiles for Li ions at points II and III, as well as for a reference sample that was immersed in the electrolyte. (Adapted with modifications from Granqvist, C.G. et al. *Electrochimica Acta*, 2018. 259: p. 1170–1182. [21]; Wen, R.-T. et al. *Nature Materials*, 2015. 14: p. 996. [58]; Arvizu, M.A. et al. *ACS Applied Materials & Interfaces*, 2015. 7(48): p. 26387–26390. [115])

hypothesis was confirmed by studying WO_3 film at a constant current density of 10^{-5} A cm^{-2} for 20 h in the "bleaching direction" [111–114]. The voltage then grew from about 2.8 to about 5.5 V, as shown in Figure 2.8b, while the optical transmittance remained high. Subsequent to this galvanostatic treatment, the open-circuit voltage (OCV) returned to its starting value of about 3.3 V, thus indicating that the pristine properties were recovered. The CV cycling data returned to those for the pristine film, as indicated in Figure 2.8a. Furthermore, optical transmittance provided further confirmation that the film had undergone regeneration, which is clear from the information in Figure 2.8c and from data on T(λ) in Figure 2.8d. Potentiostatic regeneration revealed prominent peak in the current density which was associated with a rapid change in the optical transmittance. The above results inferred that Li ion de-trapping is possible, and conclusive proof on this matter was attained by analyzing the Li content along the depth profile using time-of-flight secondary ion mass spectroscopy (ToF-SIMS) and time-of-flight elastic recoil detection analysis (ToF-ERDA) [115,116].

It was also shown that the optical transmittance modulation gets rapidly diminished during CV cycling at a certain point (point II in Figure 2.8e). The Li content was investigated and found to be much greater than in the pristine film. Galvanostatic regeneration according to the procedures delineated above resulted in abrupt onset of high transmittance after about 9 h. This could be correlated with de-trapping of Li ions as evident from the absence of Li in the ToF-ERDA data. Studies have shown that electrochemical regeneration can also be accomplished in anodically coloring EC NiO films corresponding to point III in Figure 2.8e when immersed in an electrolyte of $LiClO_4$ in propylene carbonate. The ion accumulation and ion release in this system again occurs and, interestingly, involves Li as well as Cl. Figure 2.8f shows transmittance data where the optical modulation drops during 500 electrochemical cycles, which corresponds to 11 h of treatment. Potentiostatic regeneration was then performed for a period of 20 h. During this time the transmittance reached allow level, and this treatment was repeated after a resting period of 2 h. The inset data in Figure 2.8e indicate that the initial optical performance was regained. The data for NiO show that both the cathodic and anodic parts of an EC device can be regenerated. However, work demonstrating the complete regeneration of an EC device is required. It should be noted that the physio-chemistry for the rejuvenation processes still are not well understood and more studies in this regard have been undertaken [117].

2.7 ROLE OF NANOSTRUCTURES

Nanostructuring of an electrode material in EC devices is important, and it involves several length scales. Most of EC oxides are built from octahedral structural units arranged with various degrees of corner-sharing and edge-sharing. This structure is quite beneficial since inter-octahedral spaces can facilitate ion transport. For stoichiometric WO_3, the simplest structure comprises corner-sharing octahedra, each having a centrally positioned W atom surrounded by six O atoms. However, at normal temperature and pressure, a monoclinic or triclinic structure is applicable for bulk-like WO_3 rather than a simple cubic structure. The former structures are more suitable for ion transport than the cubic ones. This is because the distances between the octahedral units are larger, which can facilitate intra-structure ion movements. Thin films, nanorods, and nanowires based on WO compositions show hexagonal structures that are readily formed, and this structure is considered to be even better for ion transport. Empirical information on nanostructures in WO thin films has been reported for evaporated and sputter deposition techniques and is being investigated by X-ray spectroscopy. In the case of evaporated thin films, agglomerated structures based on hexagonal-like units have been reported. These units grew in size and interconnectivity during film deposition onto substrates at higher temperatures [118,119].

These results concur with the fact that trimeric W_3O_9 molecules form during the evaporation thin film process, and it also has the lowest-energy structures of $(WO_3)_q$ clusters. Sputtered films apparently show higher degree of disorderliness where detailed structural models have been derived from X-ray absorption fine structure data using Monte-Carlo modelling. These films can be explained as a combination of corner-sharing and a minor amount of edge-sharing polyhedra, with five-fold and a small number of four-fold coordinated W atoms present in addition to the six-fold ones. Although EC films are described as "amorphous," some local order prevails on the level of the polyhedral structural units. This happens even when the films are deposited onto substrates at room temperature. It is observed that local order can be decreased under Li ion insertion. A certain degree of ion transfer is possible in the structures discussed above. Nevertheless, it is important to produce films with a sufficient degree of porosity to facilitate ion mobility. Many deposition techniques can be customized to produce thin films with the desired properties. Sputtering is usually employed for the manufacturing of EC glazing. For these types of deposition techniques, it is important to consider "zone diagrams," which describe characteristic structural features. These structural features can be corroborated to the deposition conditions such as the pressure of the sputter plasma and the substrate temperature in relation to the melting point of the deposited material. High pressure and low substrate temperature lead to films with columnar features and are often referred to as "zone 1" structures. Sputtered WO films under such conditions exhibit porosities of 20%–30%. Greater porosity can be achieved by the oblique-angle deposition method. Here, a large angle between the direction of the sputtered species and the substrate's surface is maintained. This approach has been demonstrated for WO-based EC films where greater porosity was obtained. A discussion on optimized nanostructures for EC thin films also includes the influence of the surface of the film. Recent works on anodic coloring NiO in Li-ion conducting electrolytes have pointed to the importance of surface features, such as deposition-dependent exposed crystal facets [120–122].

Recent studies have shown that nanostructured materials offer great potential to significantly improve the EC performance.

Furthermore, EC film processing with nanostructured materials is compatible with low-temperature processing techniques, such as solution deposition and low-cost manufacturing techniques like slot-die coating, roll-to-roll printing, flexographic printing, inkjet printing, and lamination. This can significantly improve the throughput rate and reduce operational costs, thereby opening new avenues to mass-produce smart windows similar to the printing of newspapers or banknotes, providing cheap and sustainable smart windows for global use. Toward this objective, the following section covers the recent trends and developments in zero-, one-, two-, and three-dimensional hierarchical nanostructured and ordered macroporous EC materials for advanced applications. Also, Figure 2.9 shows the scanning electron microscopy (SEM) and transmission electron microscopy images of typical 0-D, 1-D, 2-D, and 3-D nanostructures. Each of these systems is discussed in detail in subsequent sections [123].

2.7.1 Zero-Dimensional Nanostructured EC Materials

Zero-dimensional (0-D) nanostructured EC materials such as quantum dot, nanocrystal, nanoparticle arrays, and hybrid nanoparticles can be processed using different techniques. 0-D nanostructured EC materials are scalable and can be used for processing large area films with techniques like slot-die coating, inkjet printing, spray coating, spin coating, roll-to-roll printing, flexographic printing, and lamination. 0-D WO_3 particles with sizes ranging from 10–40 nm were processed using electrodeposition techniques where solutions containing $Na_2WO_4 \cdot 2H_2O$ were deposited and heat treated on flexible silver grids/PEDOT:PSS transparent conductors. The processed WO_3 films showed changes in color reversibly from transparent to dark blue with a small AC field. The obtained WO_3 nanoparticles showed an optical modulation of 81.9% at 633 nm, a high coloration efficiency (CE) value of 124.5 cm^2 C^{-1}, and fast switching speed of 2.8 and 1.9 s for bleaching and coloration processes, respectively. Moreover, good

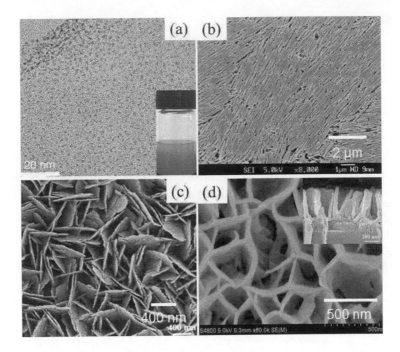

FIGURE 2.9 (a) TEM images of tungsten oxide quantum dots, (b) SEM images of the WO_3 nanorods, (c) SEM images of WO_3 nanosheet film grown on FTO substrate without any capping agent assistance. (d) SEM images of porous NiO nanosheet film. (Adapted with modification from Cai, G. et al. *Advanced Sustainable Systems*, 2017. 1(12): p. 1700074. [123]; Cai, G. et al. *Nano Energy*, 2015. 12: p. 258–267. [130]; Zhang, J. et al. *Electrochimica Acta*, 2013. 99: p. 1–8. [157])

cycling stability with transmittance modulation retention of 79.1% after 1,000 cycles was achieved. WO_3 crystalline nanoparticles synthesized through a "nano-to-nano" electrodeposition approach exhibited a particle size of 50–80 nm. Here, the electrodeposition solution was 5 wt% nanocrystalline WO_3 dispersed in water without any additive. The electrodeposited film showed about 92% optical modulation, 9 and 15 s for coloring and bleaching, respectively, and 76% optical modulation after 1,000 cycles. To enhance the EC performance, hybrid WO_3 nanoparticles were also fabricated by electrodeposition techniques. This was done using a one-pot sequential electrochemical deposition approach where the deposited hybrid thin film comprised stacked poly(3,4-ethylenedioxythiophene):poly(4-styrenesulfonic-acid) (PEDOT:PSS) and WO_3 nanoparticles. This hybrid thin film exhibited highly improved optical modulation and stability compared with pristine PEDOT: PSS or WO_3 films. 0-D EC nanomaterials like NiO and Prussian blue can be processed using electrodeposition techniques exhibiting ultrafast switching speed.

In addition to electrochemical deposition techniques, sol-gel is another common method for EC nanoparticle processing. These nanoparticles were coated on transparent conductors by inkjet printing or a spray coating technique on a given area. The film exhibited large optical modulation and fast switching speed. Solvothermal is another simple and low-cost technique for producing uniform EC nanoparticles. Uniform NiO nanoparticles were reportedly synthesized on different substrates at 200°C for 24 h with nickel acetylacetonate dispersed in tert-butanol as precursor. The color of the NiO EC materials showed reversible transformation from transparent to brown color with an applied AC field. These NiO particles showed an optical modulation of 63.6% at 550 nm, a CE of 42.8 cm^2 C^{-1} at 550 nm with 5,000 cycles [124–127].

A WO_3 quantum dot with an average crystalline size of 1.6 nm was prepared using a colloidal process. The WO_3 quantum dot was coated on fluorine-doped tin oxide (FTO) glass, and they showed excellent EC performance including fast switching speed (within 1 s), high CE (154 cm^2 C^{-1}), and an optical modulation of 85% at 633 nm. Composite nanocrystals were produced by introducing tin-doped indium oxide (Sn-ITO) nanocrystals into niobium oxide glass, and its capability of controlling visible light and near-infrared transmittance individually was realized. Such a double-band film system provides more bandwidth for the use of smart windows where the user can opt for the bright mode, which admits both NIR and visible light. The cool mode in this system can selectively block NIR light, or the opaque mode can block both NIR and visible light depending on the outside weather conditions [128–131].

2.7.2 One-Dimensional Nanostructured EC Materials

One-dimensional (1-D) nanostructured materials, such as nanowires, nanorods, nanobelts, nanotubes, and nanobundles, have incited a great interest due to their importance in the development of high performance EC devices. They play a significant role due to their high aspect ratio with nanoscale dimensions. Crystalline WO_3 nanorods with diameters about 100 nm and lengths about 2 μm were processed via a simple hydrothermal technique using NaCl as a capping agent and Na_2WO_4 solution as precursor [132,133]. After coating these WO_3 nanorods onto ITO glass, it exhibited EC phenomenon in both organic (lithium perchlorate ($LiClO_4$) in propylene carbonate (PC)) and aqueous (H_2SO_4) electrolytes. These films showed an optical modulation of 66% at −3.0 V and exhibited >3,000 cycles in organic electrolyte and 33.9% of optical modulation at 632.8 nm at −1.0 V in H_2SO_4 aqueous electrolyte. Hexagonal WO_3 nanowire array film on FTO-coated glass was prepared using $(NH_4)_2SO_4$ as capping agent where the length and diameter of nanowires obtained was about 1.5 μm and 20–40 nm, respectively and had a BET surface area of 116.5 m^2 g^{-1} [134]. These WO_3 nanowires showed an optical modulation of 58% at 633 nm and CE of 102.8 cm^2 C^{-1} in polycarbonate electrolyte. The morphology and crystallinity of the nanowires can be tailored by tuning the experimental conditions like temperature, capping agent, precursor concentration, processing pressure, etc. It was also observed that Ti and Mo doping can result in significant changes in morphology and crystallinity of the WO_3 nanowires and affect the EC performance [135,136].

It was also noted that the urea content in precursor solution played a significant role in determining the morphology of the WO_3 nanostructures. In addition to the hydrothermal method, WO_3 nanowires can be processed using solvothermal, electrospinning, or electrophoretic deposition techniques. 1-D TiO_2 have also been extensively reported for EC devices. But, the optical modulation of pure TiO_2 is small, which has limited its applications. Hence, doping of other nanostructures to improve its EC performance is a smart approach. Remarkable enhanced EC performance in terms of optical modulation, stability, and ample CE was obtained when WO_3 nanoparticles and PANI were electro-deposited on the surface of TiO_2 nanorods [137,138]. The enhanced EC properties were attributed to the porous core/shell structure, which facilitates the ion diffusion and charge-transfer. It was demonstrated that the optical modulation of TiO_2 nanotubes can be significantly enhanced by compositing with WO_3 nanostructures. It was also observed that the TiO_2/WO_3 core/shell nanowire structure showed high optical modulation and CE, which was attributed to the porous composite nanowire structure that could facilitate an improved proton intercalation capacity. Coating of MoO_3 layer on TiO_2 nanotubes system showed the optical density to increase by more than four times compared to bare TiO_2 nanotubes. Another extensively studied EC material is 1-D V_2O_5, which exhibits a color transition of green-blue and orange on applying negative and positive potentials, respectively. Xiong et al. have shown that high aspect ratio silver vanadium oxide and V_2O_5 nanowires with lengths over 30 μm and 10–20 nm diameter can be produced by hydrothermal technique. The EC device fabricated from these nanowires showed a switching time of 0.2 s for color transition from the green to the red-brown state with optical modulation of 60%. The EC performances of the V_2O_5 nanoribbons prepared by electrodeposition were enhanced by doping with Ti. This doping resulted in higher optical modulation (51.1%), higher CE (95.7 cm^2 C^{-1}) at 415 nm, and faster switching speed compared with the pristine V_2O_5. To further improve the EC performance, nanobelt-membrane hybrid structured V_2O_5 was prepared by a hydrothermal technique, which revealed the width of around 20–40 nm and displayed high optical modulation of 62% at 700 nm. In this study, it was also noted that the stability of the EC device could be further enhanced with linear polyethylenimine (LPEI) surface treatment on the transparent conductor [132,139].

Transparent 1-D NiO nanorods were produced on a conducting ITO thin film via hot-filament metal-oxide vapor deposition technique. The film showed nanorods grown within a square micrometer, exhibiting a length and width of 500 and 100 nm, respectively and an optical modulation of about 60%. These 1-D NiO nanorods also displayed stable and reversible coloration and bleaching cycles and fast coloration and bleaching speed, as well as high coloration efficiency of 43.3 cm^2 C^{-1}. Other 1-D EC nanomaterials such as Co_3O_4 nanowires and PANI nanowires show outstanding performances and can be processed by using different techniques [123].

2.7.3 Two-Dimensional Nanostructured EC Materials

Due to their unique shape-dependent characteristics, two-dimensional (2-D) nanostructured EC materials, such as nanoflakes, nanosheets, nanowalls, and nanoplates are gaining a lot of attention. Porous WO_3 films with 2-D flake nanostructures were prepared by pulsed electrochemical deposition method.

The thickness of the WO_3 flake was observed to be about 25 nm and was interconnected with each other [140,141]. This nanostructured WO_3 film displayed an optical modulation of 97.7% at 633 nm, fast switching speed of 6 and 2.7 s for coloration and bleaching process, respectively, and CE of 118.3 cm^2 C^{-1} with excellent cycling stability. Results of Wang et al.'s studies show 2-D crystalline WO_3 nanosheets being prepared on FTO coated glass via the layer-by-layer (LBL) technique. In this study, the thickness of the nanosheets was found to be about 30 nm with lateral sizes in the range of 300–500 nm. This nanosheet film displayed an optical modulation of 48.5% at 800 nm and a CE of 32 cm^2 C^{-1} [142–145].

One of the easiest ways to fabricate 2-D WO_3 EC nanomaterials is via the hydrothermal technique. Hydrothermal processing gives the flexibility of tuning the morphology and sizes depending on the requirement. For instance, Jiao et al. synthesized WO_3 nanoplates on FTO glass by the hydrothermal approach where they tuned the morphology by adding different capping agents such as Na_2SO_4, $(NH_4)_2SO_4$, or CH_3COONH_4 in the precursor. These nanoplate films exhibited high CE of 112.7 cm^2 C^{-1} with fast switching speed of 4.3/1.4 s for coloration and bleached process, respectively. Studies by Cai et al. have illustrated that the WO_3 nanosheets can be processed by adjusting the pH value of the precursor without using a template and capping agent. The thickness of the WO_3 nanosheets was found to be 10–15 nm, and it showed an optical modulation in both visible (62% at 633 nm) and NIR range (67% at 2000 nm), fast switching speed of 5.2 and 2.2 s for coloration and bleaching process, respectively, and retention of 95.4% of optical transmittance even after 3,000 cycles [146].

2-D structured NiO nanosheets have gained considerable interest because of their wide optical modulation range, high CE, and low material cost. Tu and team had reportedly synthesized NiO nanosheets on ITO glass by a combination of a chemical bath deposition technique, which was followed by heat-treatment process. The NiO nanosheet film displayed an optical modulation of about 82% at 550 nm with a switching speed of 8/10 s for coloration/bleached process, respectively, and CE of 42 cm^2 C^{-1}. It is anticipated that the EC performances such as switching speed, CE, and stability can be further improved by the addition of graphene or doping Co element in the NiO nanosheet. Similarly, 2-D structured NiO nanosheets, nanowalls, or nanoflakes can be prepared using other synthesis techniques like electrodeposition and hydrothermal methods [147–149].

2.7.4 Three-Dimensional Hierarchical Nanostructured EC Materials

Three-dimensional (3-D) hierarchical nanostructures are an ordered arrangement of low dimensional nanomaterials, which form the building blocks from the nanometer to the macroscopic scale. The 0-D, 1-D, and 2-D nanostructure elements in the 3-D material form interconnected interfaces in repetitive assemblies. EC materials showing 3-D hierarchical morphologies include nanotree arrays, nanocluster, nanoflowers, gyroid, and urchins [149,150]. These structures display large specific surface area, well-interconnected pores, and other superior characteristics over their bulk counterparts. The solvothermal method (without any template) was used to produce a variety of 3-D hierarchical nanostructures of WO_3 exhibiting nanocluster, nanotree, and nanowire arrays on FTO-coated glass through an alcoholysis reaction process between $W(CO)_6$ and ethanol. The thickness of the obtained nanostructured array was found to be about 1.1 μm [151]. The WO_3 nanotree arrays exhibited high optical modulation in visible (66.5% at 633 nm), NIR (73.8% at 2 μm), and mid-infrared ranges (57.7% at 8 μm) with a fast switching speed of 4.6 and 3.6 s for coloration and bleached processes, respectively. The CE of 126 $cm^2\ C^{-1}$ at 633 nm and cycling stability (optical modulation retention of about 80% after 4,500 cycles) was attained. Steiner et al.'s studies have shown that synthesis of V_2O_5 and NiO in a 3-D periodic interconnected gyroid structure on the nanoscale leads to remarkable improvement of EC performance such as enhanced ions intercalation, high coloration contrast, and fast switching speeds [152]. Nanourchin structured tungsten oxide comprising the $W_{18}O_{49}$ nanowires added to the sphere shell was synthesized by the solvothermal method. The nanourchin-like EC film exhibited fast switching speed and high CE of 132 $cm^2\ C^{-1}$. These films also showed good durability in the acidic electrolyte [153]. Xiao et al. also prepared a similar sphere shell structure of WO_3 (comprising hexagonal-phase nanosheets) by the one-pot hydrothermal technique with the assistance of Na_2SO_4. Here, Na_2SO_4 played the role of both stabilizer and directing agent which facilitated the generation of a metastable hexagonal phase and also assisted nanosheet assembly [154]. This hexagonal-phase WO_3 hierarchical structure exhibited an optical modulation of about 33% at 700 nm with a switching speed of 90 and 60 s for coloration and bleached processes, respectively, in organic $LiClO_4$ electrolyte at a bias voltage of ±3.0 V.

NiO showing dandelion flower-like structures was also studied for its EC performance. The size of these hierarchical structures was found to be between 2 and 3 μm, comprising nanoflakes with an average thickness of 35–40 nm. The dandelion flower-like films of NiO showed excellent EC behavior. They displayed high optical modulation of 68.09%, high CE of 88 cm^2 C^{-1}at 555 nm, and fast switching speed of 5.84 and 4.43 s for coloration and bleached processes, respectively [155].

2.7.5 Three-Dimensional Ordered Macroporous EC Materials

Besides the nanostructured EC materials, three-dimensional (3-D) ordered macroporous EC materials have gained considerable attention due to their outstanding EC performance. For preparing 3-D ordered macroporous materials, mono dispersed polystyrene latex spheres are often used as templates. Tu and team in their studies have reported the preparation of macroporous WO$_3$ and Co$_3$O$_4$ EC films by using monolayer polystyrene sphere templates [156–158]. After removal of the polystyrene sphere templates, these macroporous films displayed inter-connected network of close-packed micro-bowl arrays having an average size of 600 nm. The macroporous EC films showed superior EC properties with higher CE, faster switching speed, and higher transmittance modulation compared to dense EC film. Macroporous NiO film using similar template methods was also demonstrated by Yuan and team who observed an optical modulation of about 76% at 550 nm, fast switching speed (3 and 6 s for coloration and bleached processes, respectively), high CE of 41 cm^2 C^{-1}, and reasonably good stability. Li and team also processed 3-D ordered macroporous structured WO$_3$ and V$_2$O$_5$ EC films using multilayer PS spheres templates. Significant improvement in EC performance was obtained in this multilayer structure, which was attributed to its porous nature that facilitated the ion diffusion [159–161].

2.8 PERSPECTIVES

EC glazing has made significant commercial progress since its beginnings in 1984. Today, these glazings are manufactured by at least four companies and are installed in buildings, especially in Europe and the United States. These glazing are capable of yielding energy efficiency along with indoor comfort and other amenities. The key to successful EC products is low-cost manufacturing and long-term durability. Thin film coatings can be employed to create light-weight, robust EC devices in the form of large thin flexible panels suitable for glass lamination and incorporation in glazing [4,162].

The reactive DC magnetron sputtering technique is apparently the preferred deposition technology. However, other options, for example based on sol-gel deposition or inkjet printing, can also be used. EC products do not rely on "critical," rare earth elements, which ensures the supply chain of raw materials for manufacturing. The electrolyte in the EC device can also be a bio-hybrid, which supports the notion that EC devices are a green technology. Although predictions about future developments are notoriously difficult, there are some perspectives of exploiting the multi-functionality of EC devices. For instance, it is possible to combine energy generation, energy storage, or light-emission with electrochromism. Another aspect of multi-functionality relates to the "dual-band" of EC devices that are capable of separately modulating the luminous region and the near-infrared region. Another likely development of EC technology is to exploit thermochromic (temperature-dependent) control of solar energy throughput, which can be added to electrochromism through VO_2-based nanoparticles in the electrolyte. This is analogous with the application of nanoparticles mentioned in previous sections. It is also possible to invoke photocatalytic remediation of indoor air with EC glazing. The temperature increase caused by solar absorption in a blackened EC device can contribute to air purification. Although the discussion in this chapter was centered on glass-based products, which can be well-established in sports stadiums, function halls, etc., there are membrane systems that are based on transparent ethylene tetrafluoroethylene (ETFE) with proven durability for many decades even under full solar irradiation. Coating ETFE with a transparent conducting material is a critical step. Recent advances in thin-film deposition techniques and heat treatment of thin films on polymer substrates indicate the possibilities that these technical challenges can be met. Flexible thin film EC membranes, therefore, stand out as one of the interesting possibilities for future innovative architecture [163–166].

A wide range of EC windows exist. But in reality, only a few have undergone the durability tests and production cost analyses associated with their targeted markets. The quality requirements for architectural glass are quite stringent and EC "limitations" only relate to this application. The EC device will be deployed into standard dual pane insulated glass (IG) units on the first inner surface from the exterior. To have a double pane comprising an EC element in either the laminated or solution phase would be unacceptable to the construction industry as it increases both the cost and weight of the glass [167–170]. Thus, the thin film approach, which can be independent of ambient humidity for its operation, would seem ideal in this case. Since the spacing between the glass panes (insulated glass units) are evacuated or filled with Ar, Li systems without airtight protective layers can be processed. Systems that can provide reflective modulation based on requirements would have the advantage of reduced thermal stress in large area architectural windows. But this has not been successful so far. Widely used $c\text{-Li}_x\text{WO}_3$ still has a high absorption component. Unlike architectural glass, window products employed in automotive applications have stringent safety requirements. Thus, laminated structures employed must have quality parameters consistent with the standard safety glass construction and shatter characteristics. EC coatings on curved glass surfaces is a big challenge since these films need to withstand high processing temperatures used in glass bending. Sunroofs needing to match curved glass surfaces may also present a challenge for solution or battery-like laminated EC devices. EC technology can have high value addition to products such as sunglasses, light dimmers, optical filters, advertising displays, welding masks, etc. These applications are less sensitive to cost and/or safety issues [130,171,172]. Laminated inorganic and conductive polymer-based EC windows can be more appealing for non-architectural applications where areal weight, rate of processing and durability are not as critical as for architectural glass. Thus, from a manufacturing standpoint of large area EC windows, it appears that solution phase structures are more economical based on the aggressive cost constraints. The pre-reduction process is not required in solution phase EC windows as the complementary components can be readily mixed at the optimal concentration. There can also be a process advantage for EC windows where they can be charged after fabrication using electrolysis of bound water, hence it may be possible to avoid a pre-reduction step. High rate dc sputtering processes can be employed to generate thin film systems of WO_3, NiO, and/or IrO_2 using the respective metal targets. Both laminated and thin film, dry Li-based windows face challenges to manufacturing due to the need of pre-reduction and the requirement for high rate thin film processing [121,135,173]. However, thin film Li based windows extend the possibility of spectral selectivity that makes them attractive as a goal for future development for architectural glass applications.

REFERENCES

1. Habib, M.A. Some aspects of electrochromism of WO films prepared by a chemical deposition method. In *33rd Annual Technical Symposium*. 1989. SPIE.
2. Lampert, C.M. and R.S. Caron-Popowich. Electron microscopy and electrochemistry of nickel oxide films for electrochromic devices produced by different techniques. In *33rd Annual Technical Symposium*. 1989. SPIE.
3. Rauh, R.D. Electrochromic windows: An overview. *Electrochimica Acta*, 1999. 44(18): p. 3165–3176.
4. Svensson, J.S.E.M. and C.G. Granqvist. Electrochromic tungsten oxide films for energy efficient windows. *Solar Energy Materials*, 1984. 11(1): p. 29–34.
5. Michot, C., D. Baril, and M. Armand. Polyimide-polyether mixed conductors as switchable materials for electrochromic devices. *Solar Energy Materials and Solar Cells*, 1995. 39(2): p. 289–299.

6. Svensson, J.S.E.M., and C.G. Granqvist. Electrochromic coatings for "smart windows." *Solar Energy Materials*, 1985. 12(6): p. 391–402.

7. Sbar, N.L., L. Podbelski, H.M. Yang, and B. Pease. Electrochromic dynamic windows for office buildings. *International Journal of Sustainable Built Environment*, 2012. 1(1): p. 125–139.

8. Thummavichai, K., Y. Xia, and Y. Zhu. Recent progress in chromogenic research of tungsten oxides towards energy-related applications. *Progress in Materials Science*, 2017. 88: p. 281–324.

9. Sala, R.L., R.H. Gonçalves, E.R. Camargo, and E.R. Leite. Thermosensitive poly(N-vinylcaprolactam) as a transmission light regulator in smart windows. *Solar Energy Materials and Solar Cells*, 2018. 186: p. 266–272.

10. Rai, V., N. Tiwari, M. Rajput, S.M. Joshi, A.C. Nguyen, and N. Mathews. Reversible electrochemical silver deposition over large areas for smart windows and information display. *Electrochimica Acta*, 2017. 255: p. 63–71.

11. Cho, S.M., S. Kim, T.-Y. Kim, C.S. Ah, J. Song, S.H. Cheon, J.Y. Kim et al. New switchable mirror device with a counter electrode based on reversible electrodeposition. *Solar Energy Materials and Solar Cells*, 2018. 179: p. 161–168.

12. Dhandayuthapani, T., R. Sivakumar, R. Ilangovan, C. Gopalakrishnan, C. Sanjeeviraja, and A. Sivanantharaja. High coloration efficiency, high reversibility and fast switching response of nebulized spray deposited anatase TiO_2 thin films for electrochromic applications. *Electrochimica Acta*, 2017. 255: p. 358–368.

13. Aste, N., M. Buzzetti, C. Del Pero, R. Fusco, D. Testa, and F. Leonforte. Visual performance of yellow, orange and red LSCs integrated in a smart window. *Energy Procedia*, 2017. 105: p. 967–972.

14. Piccolo, A., C. Marino, A. Nucara, and M. Pietrafesa. Energy performance of an electrochromic switchable glazing: Experimental and computational assessments. *Energy and Buildings*, 2018. 165: p. 390–398.

15. Cannistraro, M., M.E. Castelluccio, and D. Germanò. New sol-gel deposition technique in the smart-windows – computation of possible applications of smart-windows in buildings. *Journal of Building Engineering*, 2018. 19: p. 295–301.

16. Liu, Y., L. Sun, G. Sikha, J. Isidorsson, S. Lim, A. Anders, B. Leo Kwak, and J.G. Gordon. 2-D mathematical modeling for a large electrochromic window—Part I. *Solar Energy Materials and Solar Cells*, 2014. 120: p. 1–8.

17. You, H., and A.J. Steckl. Versatile electrowetting arrays for smart window applications-from small to large pixels on fixed and flexible substrates. *Solar Energy Materials and Solar Cells*, 2013. 117: p. 544–548.

18. Patil, R.A., R.S. Devan, Y. Liou, and Y.-R. Ma. Efficient electrochromic smart windows of one-dimensional pure brookite TiO_2 nanoneedles. *Solar Energy Materials and Solar Cells*, 2016. 147: p. 240–245.

19. Liu, Q., G. Dong, Q. Chen, J. Guo, Y. Xiao, M.-P. Delplancke-Ogletree, F. Reniers, and X. Diao. Charge-transfer kinetics and cyclic properties of inorganic all-solid-state electrochromic device with remarkably improved optical memory. *Solar Energy Materials and Solar Cells*, 2018. 174: p. 545–553.

20. Ho, K.C., D.E. Singleton, and C.B. Greenberg. The influence of terminal effect on the performance of electrochromic windows. *Journal of the Electrochemical Society*, 1990. 137(12): p. 3858–3864.

21. Granqvist, C.G., M.A. Arvizu, İ. Bayrak Pehlivan, H.Y. Qu, R.T. Wen, and G.A. Niklasson. Electrochromic materials and devices for energy efficiency and human comfort in buildings: A critical review. *Electrochimica Acta*, 2018. 259: p. 1170–1182.

22. Allen, K., K. Connelly, P. Rutherford, and Y. Wu. Smart windows—dynamic control of building energy performance. *Energy and Buildings*, 2017. 139: p. 535–546.

23. Rauh, R.D., and S.F. Cogan. Design model for electrochromic windows and application to the WO_3/IrO_2 system. *Journal of the Electrochemical Society*, 1993. 140(2): p. 378–386.

24. Kwon, S.-B., S.-J. Lee, D.-S. Yoon, H.-S. Yoo, and B.-Y. Lee. Transmittance variable liquid crystal modes with a specific gray off-state for low power consumption smart windows. *Journal of Molecular Liquids*, 2018.

25. Chen, D., R. Zhang, R. Wang, L.D. Negro, and S.D. Minteer. Gold nanofiber-based electrodes for plasmon-enhanced electrocatalysis. *Journal of the Electrochemical Society*, 2016. 163(14): p. H1132–H1135.

26. Dydek, E.V., M.V. Petersen, D.G. Nocera, and K.F. Jensen. Electrode placement and fluid flow rates in microfluidic electrochemical devices. *Journal of the Electrochemical Society*, 2012. 159(11): p. H853–H856.

27. Arıcan, D., A. Aktaş, H. Kantekin, and A. Koca. Electrochromism of electropolymerized phthalocyanine-tetrahydroquinoline dyads. *Journal of the Electrochemical Society*, 2014. 161(10): p. H670–H676.

28. Li, W., L. Chen, Y. Pan, S. Yan, Y. Dai, J. Liu, Y. Yu et al. Electrochromic properties of polymers/copolymers via electrochemical polymerization based on star-shaped thiophene derivatives with different central cores. *Journal of the Electrochemical Society*, 2017. 164(4): p. E84–E89.

29. Granqvist, C.G. *Handbook of Electrochromic Materials*. 1995, Amsterdam: Elsevier Science B.V.

30. Deb, S.K. Optical and photoelectric properties and colour centres in thin films of tungsten oxide. *The Philosophical Magazine: A Journal of Theoretical Experimental and Applied Physics*, 1973. 27(4): p. 801–822.

31. Agrawal, A., J.P. Cronin, and R. Zhang. Review of solid state electrochromic coatings produced using sol-gel techniques. *Solar Energy Materials and Solar Cells*, 1993. 31(1): p. 9–21.

32. Cogan, S.F., T.D. Plante, R.S. McFadden, and R.D. Rauh. Solar modulation in a-WO_3/a-IrO_2 and c-$KxWO_3$ + (x/2)/a-IrO_2 complementary electrochromic windows. *Solar Energy Materials*, 1987. 16(5): p. 371–382.

33. Yuan, G., C. Hua, L. Huang, C. Defranoux, P. Basa, Y. Liu, C. Song, and G. Han. Optical characterization of the coloration process in electrochromic amorphous and crystalline WO_3 films by spectroscopic ellipsometry. *Applied Surface Science*, 2017. 421: p. 630–635.

34. Ge, C., M. Wang, S. Hussain, Z. Xu, G. Liu, and G. Qiao. Electron transport and electrochromic properties of sol-gel WO_3 thin films: Effect of crystallinity. *Thin Solid Films*, 2018. 653: p. 119–125.

35. Usha, N., R. Sivakumar, and C. Sanjeeviraja. Structural, optical and electrochromic properties of Nb_2O_5:MoO_3 (95:5, 90:10, and 85:15) thin films prepared by RF magnetron sputtering technique. *Materials Letters*, 2018. 229: p. 189–192.

36. Tang, C.-J., J.-M. Ye, Y.-T. Yang, and J.-L. He. Large-area flexible monolithic ITO/WO_3/Nb_2O_5/NiVOχ/ITO electrochromic devices prepared by using magnetron sputter deposition. *Optical Materials*, 2016. 55: p. 83–89.

37. Taunier, S., C. Guery, and J.M. Tarascon. Design and characterization of a three-electrode electrochromic device, based on the system WO_3/IrO_2. *Electrochimica Acta*, 1999. 44(18): p. 3219–3225.

38. Rauh, D. Design and fabrication of electrochromic light modulators. *Solar Energy Materials and Solar Cells*, 1995. 39(2): p. 145–154.

39. Li, C., J.H. Hsieh, T.Y. Su, and P.L. Wu. Experimental study on property and electrochromic function of stacked WO_3/Ta_2O_5/NiO films by sputtering. *Thin Solid Films*, 2018. 660: p. 373–379.

40. Che, X., Z. Wu, G. Dong, X. Diao, Y. Zhou, J. Guo, D. Dong, and M. Wang. Properties of all-thin-film glass/ITO/WO$_3$:H/Ta$_2$O$_5$/NiO$_x$/ ITO electrochromic devices prepared by magnetron sputtering. *Thin Solid Films*, 2018. 662: p. 6–12.

41. Tian, Y., Z. Li, S. Dou, X. Zhang, J. Zhang, L. Zhang, L. Wang, X. Zhao, and Y. Li. Facile preparation of aligned NiO nanotube arrays for electrochromic application. *Surface and Coatings Technology*, 2018. 337: p. 63–67.

42. Zrikem, K., G. Song, A.A. Aghzzaf, M. Amjoud, D. Mezzane, and A. Rougier. UV treatment for enhanced electrochromic properties of spin coated NiO thin films. *Superlattices and Microstructures*, 2018.

43. Mathew, J.G.H., S.P. Sapers, M.J. Cumbo, N.A. O'Brien, R.B. Sargent, V.P. Raksha, R.B. Lahaderne, and B.P. Hichwa. Large area electrochromics for architectural applications. *Journal of Non-Crystalline Solids*, 1997. 218: p. 342–346.

44. Sone, Y., A. Kishimoto, T. Kudo, and K. Ikeda. Reversible electrochromic performance of Prussian blue coated with proton conductive Ta$_2$O$_5$·nH$_2$O film. *Solid State Ionics*, 1996. 83(1): p. 135–143.

45. Choe, H.S., B.G. Carroll, D.M. Pasquariello, and K.M. Abraham. Characterization of some polyacrylonitrile-based electrolytes. *Chemistry of Materials*, 1997. 9(1): p. 369–379.

46. Julien, C., and G.-A. Nazri. *Solid State Batteries: Materials Design and Optimization*. 1994, Boston, MA: Kluwer Academic Publishers.

47. Cogan, S.F., R.D. Rauh, J.D. Klein, N.M. Nguyen, R.B. Jones, and T.D. Plante. Variable transmittance coatings using electrochromic lithium chromate and amorphous WO$_3$ thin films. *Journal of the Electrochemical Society*, 1997. 144(3): p. 956–960.

48. Leventis, N., and Y.C. Chung. Complementary surface confined polymer electrochromic materials, systems, and methods of fabrication therefor. 1995.

49. Kamimori, T., M. Mizuhashi, and J. Nagai. Electro-optical device and electro-optical light controlling device. 1984.

50. Goldner, R.B., G. Seward, K. Wong, T. Haas, G.H. Foley, R. Chapman, and S. Schulz. Completely solid lithiated smart windows. *Solar Energy Materials*, 1989. 19(1): p. 17–26.

51. Goldner, R.B., P. Norton, K. Wong, G. Foley, E.L. Goldner, G. Seward, and R. Chapman. Further evidence for free electrons as dominating the behavior of electrochromic polycrystalline WO$_3$ films. *Applied Physics Letters*, 1985. 47(5): p. 536–538.

52. Cogan, S.F., T.D. Plante, M.A. Parker, and R.D. Rauh. Free-electron electrochromic modulation in crystalline LixWO₃. *Journal of Applied Physics*, 1986. 60(8): p. 2735–2738.
53. Goldner, R.B., D.H. Mendelsohn, J. Alexander, W.R. Henderson, D. Fitzpatrick, T.E. Haas, and H.H. Sample. High near-infrared reflectivity modulation with polycrystalline electrochromic WO₃ films. *Applied Physics Letters*, 1983. 43(12): p. 1093–1095.
54. Hjelm, A., C.G. Granqvist, and J.M. Wills. Electronic structure and optical properties of WO₃, LiWO₃, NaWO₃, and HWO₃. *Physical Review B*, 1996. 54(4): p. 2436–2445.
55. de Wijs, G.A., and R.A. de Groot. Structure and electronic properties of amorphous WO₃. *Physical Review B*, 1999. 60(24): p. 16463–16474.
56. Bondarenko, N., O. Eriksson, and N.V. Skorodumova. Polaron mobility in oxygen-deficient and lithium-doped tungsten trioxide. *Physical Review B*, 2015. 92(16): p. 165119.
57. Hamdi, H., E.K.H. Salje, P. Ghosez, and E. Bousquet. First-principles reinvestigation of bulk WO₃. *Physical Review B*, 2016. 94(24): p. 245124.
58. Wen, R.-T., C.G. Granqvist, and G.A. Niklasson. Eliminating degradation and uncovering ion-trapping dynamics in electrochromic WO₃ thin films. *Nature Materials*, 2015. 14: p. 996.
59. Berggren, L., and G.A. Niklasson. Optical charge transfer absorption in lithium-intercalated tungsten oxide thin films. *Applied Physics Letters*, 2006. 88(8): p. 081906.
60. Wen, R.-T., C.G. Granqvist, and G.A. Niklasson. Anodic electrochromism for energy-efficient windows: Cation/anion-based surface processes and effects of crystal facets in nickel oxide thin films. *Advanced Functional Materials*, 2015. 25(22): p. 3359–3370.
61. Avendaño, E., A. Azens, G.A. Niklasson, and C.G. Granqvist. Proton diffusion and electrochromism in hydrated NiO y and Ni1 − x V x O y thin films. *Journal of the Electrochemical Society*, 2005. 152(12): p. F203–F212.
62. Goodenough, J.B. Metallic oxides. *Progress in Solid State Chemistry*, 1971. 5: p. 145–399.
63. Gavrilyuk, A., U. Tritthart, and W. Gey. Photoinjection of hydrogen and the nature of a giant shift of the fundamental absorption edge in highly disordered V₂O₅ films. *Physical Chemistry Chemical Physics*, 2011. 13(20): p. 9490–9497.
64. Talledo, A., and C.G. Granqvist. Electrochromic vanadium-pentoxide–based films: Structural, electrochemical, and optical properties. *Journal of Applied Physics*, 1995. 77(9): p. 4655–4666.

65. Schirmer, O.F., V. Wittwer, G. Baur, and G. Brandt. Dependence of WO_3 electrochromic absorption on crystallinity. *Journal of the Electrochemical Society*, 1977. 124(5): p. 749–753.

66. Ederth, J., A. Hoel, G.A. Niklasson, and C.G. Granqvist. Small polaron formation in porous WO_3 − x nanoparticle films. *Journal of Applied Physics*, 2004. 96(10): p. 5722–5726.

67. Zhang, J.G., D.K. Benson, C.E. Tracy, S.K. Deb, A.W. Czanderna, and C. Bechinger. Chromic mechanism in amorphous WO_3 films. *Journal of the Electrochemical Society*, 1997. 144(6): p. 2022–2026.

68. Darmawi, S., S. Burkhardt, T. Leichtweiss, D.A. Weber, S. Wenzel, J. Janek, M.T. Elm, and P.J. Klar. Correlation of electrochromic properties and oxidation states in nanocrystalline tungsten trioxide. *Physical Chemistry Chemical Physics*, 2015. 17(24): p. 15903–15911.

69. Reik, H.G., and D. Heese. Frequency dependence of the electrical conductivity of small polarons for high and low temperatures. *Journal of Physics and Chemistry of Solids*, 1967. 28(4): p. 581–596.

70. Granqvist, C.G. Electrochromics for smart windows: Oxide-based thin films and devices. *Thin Solid Films*, 2014. 564: p. 1–38.

71. Wen, R.-T., G.A. Niklasson, and C.G. Granqvist. Strongly improved electrochemical cycling durability by adding iridium to electrochromic nickel oxide films. *ACS Applied Materials & Interfaces*, 2015. 7(18): p. 9319–9322.

72. Arvizu, M.A., C.A. Triana, B.I. Stefanov, C.G. Granqvist, and G.A. Niklasson. Electrochromism in sputter-deposited W–Ti oxide films: Durability enhancement due to Ti. *Solar Energy Materials and Solar Cells*, 2014. 125: p. 184–189.

73. Gillaspie, D., A. Norman, C.E. Tracy, J.R. Pitts, S.-H. Lee, and A. Dillon. Nanocomposite counter electrode materials for electrochromic windows. *Journal of the Electrochemical Society*, 2010. 157(3): p. H328–H331.

74. Lin, F., D. Nordlund, T.-C. Weng, R.G. Moore, D.T. Gillaspie, A.C. Dillon, R.M. Richards, and C. Engtrakul. Hole doping in al-containing nickel oxide materials to improve electrochromic performance. *ACS Applied Materials & Interfaces*, 2013. 5(2): p. 301–309.

75. Lin, F., D. Nordlund, T.-C. Weng, D. Sokaras, K.M. Jones, R.B. Reed, D.T. Gillaspie et al., Origin of electrochromism in high-performing nanocomposite nickel oxide. *ACS Applied Materials & Interfaces*, 2013. 5(9): p. 3643–3649.

76. Zhou, J., G. Luo, Y. Wei, J. Zheng, and C. Xu. Enhanced electrochromic performances and cycle stability of NiO-based thin films via Li–Ti co-doping prepared by sol–gel method. *Electrochimica Acta*, 2015. 186: p. 182–191.

77. Cha, I.Y., S.H. Park, J.W. Lim, S.J. Yoo, and Y.-E. Sung. The activation process through a bimodal transmittance state for improving electrochromic performance of nickel oxide thin film. *Solar Energy Materials and Solar Cells*, 2013. 108: p. 22–26.

78. Sapp, S.A., G.A. Sotzing, and J.R. Reynolds. High contrast ratio and fast-switching dual polymer electrochromic devices. *Chemistry of Materials*, 1998. 10(8): p. 2101–2108.

79. Leventis, N., M. Chen, A.I. Liapis, J.W. Johnson, and A. Jain. Characterization of 3 × 3 matrix arrays of solution-phase electrochromic cells. *Journal of the Electrochemical Society*, 1998. 145(4): p. L55–L58.

80. Byker, H.J. Single-compartment, self-erasing, solution-phase electrochromic devices, solutions for use therein, and uses thereof. 1990.

81. Goldner, R.B., F.O. Arntz, K. Dickson, M.A. Goldner, T.E. Haas, T.Y. Liu, S. Slaven, G. Wei, K.K. Wong, and P. Zerigian. Some lessons learned from research on a thin film electrochromic window. *Solid State Ionics*, 1994. 70-71: p. 613–618.

82. Dickens, P.G., and S.A. Kay. Thermochemistry of the cubic and hexagonal lithium tungsten bronze phases LixWO₃. *Solid State Ionics*, 1983. 8(4): p. 291–295.

83. Ceder, G., and M.K. Aydinol. The electrochemical stability of lithium-metal oxides against metal reduction. *Solid State Ionics*, 1998. 109(1): p. 151–157.

84. Cogan, S.F., E.J. Anderson, T.D. Plante, and R.D. Rauh. Electrochemical investigation of electrochromism in transparent conductive oxides. *Applied Optics*, 1985. 24(15): p. 2282–2283.

85. Kishimoto, A., T. Kudo, and T. Nanba. Amorphous tantalum and niobium oxide proton conductors derived from respective peroxo polyacids. *Solid State Ionics*, 1992. 53–56: p. 993–997.

86. Zeller, H.R., and H.U. Beyeler. Electrochromism and local order in amorphous WO₃. *Applied Physics*, 1977. 13(3): p. 231–237.

87. Li, W., W.R. McKinnon, and J.R. Dahn. Lithium intercalation from aqueous solutions. *Journal of the Electrochemical Society*, 1994. 141(9): p. 2310–2316.

88. Goldner, R.B., T.E. Haas, F.O. Arntz, S. Slaven, K.K. Wong, B. Wilkens, C. Shepard, and W. Lanford. Nuclear reaction analysis profiling as direct evidence for lithium ion mass transport in thin film "rocking-chair" structures. *Applied Physics Letters*, 1993. 62(14): p. 1699–1701.

89. Aurbach, D., M.D. Levi, E. Levi, H. Teller, B. Markovsky, G. Salitra, U. Heider, and L. Heider. Common electroanalytical behavior of li intercalation processes into graphite and transition metal oxides. *Journal of the Electrochemical Society*, 1998. 145(9): p. 3024–3034.

90. Striebel, K.A., C.Z. Deng, S.J. Wen, and E.J. Cairns. Electrochemical behavior of LiMn$_2$O$_4$ and LiCoO$_2$ thin films produced with pulsed laser deposition. *Journal of the Electrochemical Society*, 1996. 143(6): p. 1821–1827.

91. Wegener, E.J. Large volume coated glass production for architectural markets in North America. *Journal of Non-Crystalline Solids*, 1997. 218: p. 7–11.

92. Zhanga, Y., S.-H. Lee, A. Mascarenhas, and S.K. Deb. An UV photochromic memory effect in proton-based WO$_3$ electrochromic devices. *Applied Physics Letters*, 2008. 93(20): p. 203508.

93. Wang, Y., J. Kim, Z. Gao, O. Zandi, S. Heo, P. Banerjee, and D.J. Milliron. Disentangling photochromism and electrochromism by blocking hole transfer at the electrolyte interface. *Chemistry of Materials*, 2016. 28(20): p. 7198–7202.

94. Miyazaki, H., T. Matsuura, and T. Ota. Nickel oxide-based photochromic composite films. *Journal of the Ceramic Society of Japan*, 2016. 124(11): p. 1175–1177.

95. Jonsson, A., A. Roos, and E.K. Jonson. The effect on transparency and light scattering of dip coated antireflection coatings on window glass and electrochromic foil. *Solar Energy Materials and Solar Cells*, 2010. 94(6): p. 992–997.

96. Bayrak Pehlivan, İ., C.G. Granqvist, R. Marsal, P. Georén, and G.A. Niklasson. [PEI-SiO$_2$]:[LiTFSI] nanocomposite polymer electrolytes: Ion conduction and optical properties. *Solar Energy Materials and Solar Cells*, 2012. 98: p. 465–471.

97. Hamberg, I., and C.G. Granqvist. Evaporated Sn-doped In$_2$O$_3$ films: Basic optical properties and applications to energy-efficient windows. *Journal of Applied Physics*, 1986. 60(11): p. R123–R160.

98. Niklasson, G.A., C.G. Granqvist, and O. Hunderi. Effective medium models for the optical properties of inhomogeneous materials. *Applied Optics*, 1981. 20(1): p. 26–30.

99. Bishop, C. *Vacuum Deposition onto Webs, Films and Foils*. 2nd Edition. 2011. Elsevier.

100. Czanderna, A.W., D.K. Benson, G.J. Jorgensen, J.G. Zhang, C.E. Tracy, and S.K. Deb. Durability issues and service lifetime prediction of electrochromic windows for buildings applications. *Solar Energy Materials and Solar Cells*, 1999. 56(3): p. 419–436.

101. Lampert, C.M., A. Agrawal, C. Baertlien, and J. Nagai. Durability evaluation of electrochromic devices – An industry perspective. *Solar Energy Materials and Solar Cells*, 1999. 56(3): p. 449–463.

102. Nagai, J., G.D. McMeeking, and Y. Saitoh. Durability of electrochromic glazing. *Solar Energy Materials and Solar Cells*, 1999. 56(3): p. 309–319.

103. Sbar, N., M. Badding, R. Budziak, K. Cortez, L. Laby, L. Michalski, T. Ngo, S. Schulz, and K. Urbanik. Progress toward durable, cost effective electrochromic window glazings. *Solar Energy Materials and Solar Cells*, 1999. 56(3): p. 321–341.

104. Kubo, T., J. Tanimoto, M. Minami, T. Toya, Y. Nishikitani, and H. Watanabe. Performance and durability of electrochromic windows with carbon-based counter electrode and their application in the architectural and automotive fields. *Solid State Ionics*, 2003. 165(1): p. 97–104.

105. Wena, R.-T., C.G. Granqvist, and G.A. Niklasson. Cyclic voltammetry on sputter-deposited films of electrochromic Ni oxide: Power-law decay of the charge density exchange. *Applied Physics Letters*, 2014. 105(16): p. 163502.

106. Wen, R.-T., C.G. Granqvist, and G.A. Niklasson. Anodic electrochromic nickel oxide thin films: Decay of charge density upon extensive electrochemical cycling. *ChemElectroChem*, 2016. 3(2): p. 266–275.

107. Wen, R.-T., S. Malmgren, C.G. Granqvist, and G.A. Niklasson. Degradation dynamics for electrochromic WO_3 films under extended charge insertion and extraction: Unveiling physicochemical mechanisms. *ACS Applied Materials & Interfaces*, 2017. 9(14): p. 12872–12877.

108. Plonka, A. 4 Dispersive kinetics. *Annual Reports Section "C" (Physical Chemistry)*, 2001. 97(0): p. 91–147.

109. Wen, R.-T., G.A. Niklasson, and C.G. Granqvist. Eliminating electrochromic degradation in amorphous TiO_2 through Li-Ion detrapping. *ACS Applied Materials & Interfaces*, 2016. 8(9): p. 5777–5782.

110. Miguel, A.A., G.G. Claes, and A.N. Gunnar. Rejuvenation of degraded electrochromic MoO_3 thin films made by DC magnetron sputtering: Preliminary results. *Journal of Physics: Conference Series*, 2016. 764(1): p. 012009.

111. Bisquert, J. Analysis of the kinetics of ion intercalation: Ion trapping approach to solid-state relaxation processes. *Electrochimica Acta*, 2002. 47(15): p. 2435–2449.

112. Bisquert, J., and V.S. Vikhrenko. Analysis of the kinetics of ion intercalation. Two state model describing the coupling of solid state ion diffusion and ion binding processes. *Electrochimica Acta*, 2002. 47(24): p. 3977–3988.

113. Bisquert, J. Fractional diffusion in the multiple-trapping regime and revision of the equivalence with the continuous-time random walk. *Physical Review Letters*, 2003. 91(1): p. 010602.

114. Bisquert, J. Beyond the quasistatic approximation: Impedance and capacitance of an exponential distribution of traps. *Physical Review B*, 2008. 77(23): p. 235203.

115. Arvizu, M.A., R.-T. Wen, D. Primetzhofer, J.E. Klemberg-Sapieha, L. Martinu, G.A. Niklasson, and C.G. Granqvist. Galvanostatic ion detrapping rejuvenates oxide thin films. *ACS Applied Materials & Interfaces*, 2015. 7(48): p. 26387–26390.

116. Baloukas, B., M.A. Arvizu, R.-T. Wen, G.A. Niklasson, C.G. Granqvist, R. Vernhes, J.E. Klemberg-Sapieha, and L. Martinu. Galvanostatic rejuvenation of electrochromic WO_3 thin films: Ion trapping and detrapping observed by optical measurements and by time-of-flight secondary ion mass spectrometry. *ACS Applied Materials & Interfaces*, 2017. 9(20): p. 16995–17001.

117. Qu, H.-Y., D. Primetzhofer, M.A. Arvizu, Z. Qiu, U. Cindemir, C.G. Granqvist, and G.A. Niklasson. Electrochemical rejuvenation of anodically coloring electrochromic nickel oxide thin films. *ACS Applied Materials & Interfaces*, 2017. 9(49): p. 42420–42424.

118. Aliev, A.E., and H.W. Shin. Nanostructured materials for electrochromic devices. *Solid State Ionics*, 2002. 154–155: p. 425–431.

119. Baeck, S.-H., K.-S. Choi, T.F. Jaramillo, G.D. Stucky, and E.W. McFarland. Enhancement of photocatalytic and electrochromic properties of electrochemically fabricated mesoporous WO_3 thin films. *Advanced Materials*, 2003. 15(15): p. 1269–1273.

120. Lee, S.-H., R. Deshpande, P.A. Parilla, K.M. Jones, B. To, A.H. Mahan, and A.C. Dillon. Crystalline WO_3 nanoparticles for highly improved electrochromic applications. *Advanced Materials*, 2006. 18(6): p. 763–766.

121. Li, H., G. Shi, H. Wang, Q. Zhang, and Y. Li. Self-seeded growth of nest-like hydrated tungsten trioxide film directly on FTO substrate for highly enhanced electrochromic performance. *Journal of Materials Chemistry A*, 2014. 2(29): p. 11305–11310.

122. Tiwari, J.N., R.N. Tiwari, and K.S. Kim. Zero-dimensional, one-dimensional, two-dimensional and three-dimensional nanostructured materials for advanced electrochemical energy devices. *Progress in Materials Science*, 2012. 57(4): p. 724–803.

123. Cai, G., A.L.-S. Eh, L. Ji, and P.S. Lee. Recent advances in electrochromic smart fenestration. *Advanced Sustainable Systems*, 2017. 1(12): p. 1700074.

124. Cai, G., P. Darmawan, M. Cui, J. Wang, J. Chen, S. Magdassi, and P.S. Lee. Supercapacitors: Highly stable transparent conductive silver grid/PEDOT: PSS electrodes for integrated bifunctional flexible electrochromic supercapacitors (Adv. Energy Mater. 4/2016). *Advanced Energy Materials*, 2016. 6(4).

125. Liu, L., M. Layani, S. Yellinek, A. Kamyshny, H. Ling, P.S. Lee, S. Magdassi, and D. Mandler. "Nano to nano" electrodeposition of WO_3 crystalline nanoparticles for electrochromic coatings. *Journal of Materials Chemistry A*, 2014. 2(38): p. 16224–16229.

126. Ling, H., J. Lu, S. Phua, H. Liu, L. Liu, Y. Huang, D. Mandler, P.S. Lee, and X. Lu. One-pot sequential electrochemical deposition of multilayer poly(3,4-ethylenedioxythiophene):poly(4-styrenesulfonic acid)/tungsten trioxide hybrid films and their enhanced electrochromic properties. *Journal of Materials Chemistry A*, 2014. 2(8): p. 2708–2717.

127. Cai, G.F., C.D. Gu, J. Zhang, P.C. Liu, X.L. Wang, Y.H. You, and J.P. Tu. Ultra fast electrochromic switching of nanostructured NiO films electrodeposited from choline chloride-based ionic liquid. *Electrochimica Acta*, 2013. 87: p. 341–347.

128. Layani, M., P. Darmawan, W.L. Foo, L. Liu, A. Kamyshny, D. Mandler, S. Magdassi, and P.S. Lee. Nanostructured electrochromic films by inkjet printing on large area and flexible transparent silver electrodes. *Nanoscale*, 2014. 6(9): p. 4572–4576.

129. Costa, C., C. Pinheiro, I. Henriques, and C.A.T. Laia. Inkjet printing of sol–gel synthesized hydrated tungsten oxide nanoparticles for flexible electrochromic devices. *ACS Applied Materials & Interfaces*, 2012. 4(3): p. 1330–1340.

130. Cai, G., X. Wang, M. Cui, P. Darmawan, J. Wang, A.L.-S. Eh, and P.S. Lee. Electrochromo-supercapacitor based on direct growth of NiO nanoparticles. *Nano Energy*, 2015. 12: p. 258–267.

131. Cong, S., Y. Tian, Q. Li, Z. Zhao, and F. Geng. Single-crystalline tungsten oxide quantum dots for fast pseudocapacitor and electrochromic applications. *Advanced Materials*, 2014. 26(25): p. 4260–4267.

132. Kang, W., C. Yan, X. Wang, C.Y. Foo, A.W. Ming Tan, K.J. Zhi Chee, and P.S. Lee. Green synthesis of nanobelt-membrane hybrid structured vanadium oxide with high electrochromic contrast. *Journal of Materials Chemistry C*, 2014. 2(24): p. 4727–4732.

133. Wang, J., E. Khoo, P.S. Lee, and J. Ma. Controlled synthesis of WO_3 nanorods and their electrochromic properties in H_2SO_4 electrolyte. 2009. 113: p. 9655–9658.

134. Zhang, J., J.P. Tu, X.H. Xia, X.L. Wang, and C.D. Gu. Hydrothermally synthesized WO_3 nanowire arrays with highly improved electrochromic performance. *Journal of Materials Chemistry*, 2011. 21(14): p. 5492–5498.

135. Zhou, D., F. Shi, D. Xie, D.H. Wang, X.H. Xia, X.L. Wang, C.D. Gu, and J.P. Tu. Bi-functional Mo-doped WO_3 nanowire array electrochromism-plus electrochemical energy storage. *Journal of Colloid and Interface Science*, 2016. 465: p. 112–120.

136. Cai, G.F., X.L. Wang, D. Zhou, J.H. Zhang, Q.Q. Xiong, C.D. Gu, and J.P. Tu. Hierarchical structure Ti-doped WO_3 film with improved electrochromism in visible-infrared region. *RSC Advances*, 2013. 3(19): p. 6896–6905.

137. Cai, G., J. Tu, D. Zhou, J. Zhang, Q. Xiong, X. Zhao, X. Wang, and C. Gu. Multicolor electrochromic film based on TiO_2@Polyaniline core/shell nanorod array. *Journal of Physical Chemistry C*, 2013. 117: p. 15967–15975.

138. Cai, G.F., D. Zhou, Q.Q. Xiong, J.H. Zhang, X.L. Wang, C.D. Gu, and J.P. Tu. Efficient electrochromic materials based on TiO_2@WO_3 core/shell nanorod arrays. *Solar Energy Materials and Solar Cells*, 2013. 117: p. 231–238.

139. Lu, Y., L. Liu, D. Mandler, and P.S. Lee. High switching speed and coloration efficiency of titanium-doped vanadium oxide thin film electrochromic devices. *Journal of Materials Chemistry C*, 2013. 1(44): p. 7380–7386.

140. Zhang, X., Y. Zhang, B. Zhao, S. Lu, H. Wang, J. Liu, and H. Yan. Improvement on optical modulation and stability of the NiO based electrochromic devices by nanocrystalline modified nanocomb hybrid structure. *RSC Advances*, 2015. 5(123): p. 101487–101493.

141. Cai, G., M. Cui, V. Kumar, P. Darmawan, J. Wang, X. Wang, A. Lee-Sie Eh, K. Qian, and P.S. Lee. Ultra-large optical modulation of electrochromic porous WO_3 film and the local monitoring of redox activity. *Chemical Science*, 2016. 7(2): p. 1373–1382.

142. Wang, K., P. Zeng, J. Zhai, and Q. Liu. Electrochromic films with a stacked structure of WO_3 nanosheets. *Electrochemistry Communications*, 2013. 26: p. 5–9.

143. Jiao, Z., X. Wang, J. Wang, L. Ke, H.V. Demir, T.W. Koh, and X.W. Sun. Efficient synthesis of plate-like crystalline hydrated tungsten trioxide thin films with highly improved electrochromic performance. *Chemical Communications*, 2012. 48(3): p. 365–367.

144. Zhihui, J., S. Xiao Wei, W. Jinmin, K. Lin, and D. Hilmi Volkan. Hydrothermally grown nanostructured WO_3 films and their electrochromic characteristics. *Journal of Physics D: Applied Physics*, 2010. 43(28): p. 285501.

145. Jiao, Z., J. Wang, L. Ke, X.W. Sun, and H.V. Demir. Morphology-tailored synthesis of tungsten trioxide (hydrate) thin films and their photocatalytic properties. *ACS Applied Materials & Interfaces*, 2011. 3(2): p. 229–236.

146. Cai, G.F., J.P. Tu, D. Zhou, L. Li, J.H. Zhang, X.L. Wang, and C.D. Gu. The direct growth of a WO_3 nanosheet array on a transparent conducting substrate for highly efficient electrochromic and electrocatalytic applications. *CrystEngComm*, 2014. 16(30): p. 6866–6872.

147. Xia, X.H., J.P. Tu, J. Zhang, X.L. Wang, W.K. Zhang, and H. Huang. Electrochromic properties of porous NiO thin films prepared by a chemical bath deposition. *Solar Energy Materials and Solar Cells*, 2008. 92(6): p. 628–633.

148. Ma, D., G. Shi, H. Wang, Q. Zhang, and Y. Li. Hierarchical NiO microflake films with high coloration efficiency, cyclic stability and low power consumption for applications in a complementary electrochromic device. *Nanoscale*, 2013. 5(11): p. 4808–4815.

149. Cao, F., G.X. Pan, X.H. Xia, P.S. Tang, and H.F. Chen. Hydrothermal-synthesized mesoporous nickel oxide nanowall arrays with enhanced electrochromic application. *Electrochimica Acta*, 2013. 111: p. 86–91.

150. Brezesinski, T., D. Fattakhova Rohlfing, S. Sallard, M. Antonietti, and B.M. Smarsly. Highly crystalline WO_3 thin films with ordered 3D mesoporosity and improved electrochromic performance. *Small*, 2006. 2(10): p. 1203–1211.

151. Cai, G.F., J.P. Tu, D. Zhou, X.L. Wang, and C.D. Gu. Growth of vertically aligned hierarchical WO_3 nano-architecture arrays on transparent conducting substrates with outstanding electrochromic performance. *Solar Energy Materials and Solar Cells*, 2014. 124: p. 103–110.

152. Wei, D., M.R.J. Scherer, C. Bower, P. Andrew, T. Ryhänen, and U. Steiner. A nanostructured electrochromic supercapacitor. *Nano Letters*, 2012. 12(4): p. 1857–1862.

153. Kim, D.H. Effects of phase and morphology on the electrochromic performance of tungsten oxide nano-urchins. *Solar Energy Materials and Solar Cells*, 2012. 107: p. 81–86.

154. Xiao, W., W. Liu, X. Mao, H. Zhu, and D. Wang. Na_2SO_4-assisted synthesis of hexagonal-phase WO_3 nanosheet assemblies with applicable electrochromic and adsorption properties. *Journal of Materials Chemistry A*, 2013. 1(4): p. 1261–1269.

155. Dalavi, D.S., R.S. Devan, R.S. Patil, Y.R. Ma, M.G. Kang, J.H. Kim, and P.S. Patil. Electrochromic properties of dandelion flower like nickel oxide thin films. *Journal of Materials Chemistry A*, 2013. 1(4): p. 1035–1039.

156. Xia, X.H., J.P. Tu, J. Zhang, X.H. Huang, X.L. Wang, and X.B. Zhao. Improved electrochromic performance of hierarchically porous Co_3O_4 array film through self-assembled colloidal crystal template. *Electrochimica Acta*, 2010. 55(3): p. 989–994.

157. Zhang, J., J.P. Tu, G.F. Cai, G.H. Du, X.L. Wang, and P.C. Liu. Enhanced electrochromic performance of highly ordered, macroporous WO_3 arrays electrodeposited using polystyrene colloidal crystals as template. *Electrochimica Acta*, 2013. 99: p. 1–8.

158. Xia, X.H., J.P. Tu, J. Zhang, J.Y. Xiang, X.L. Wang, and X.B. Zhao. Cobalt oxide ordered bowl-like array films prepared by electrodeposition through monolayer polystyrene sphere template and electrochromic properties. *ACS Applied Materials & Interfaces*, 2010. 2(1): p. 186–192.

159. Yang, L., D. Ge, J. Zhao, Y. Ding, X. Kong, and Y. Li. Improved electrochromic performance of ordered macroporous tungsten oxide films for IR electrochromic device. *Solar Energy Materials and Solar Cells*, 2012. 100: p. 251–257.

160. Tong, Z., J. Hao, K. Zhang, J. Zhao, B.-L. Su, and Y. Li. Improved electrochromic performance and lithium diffusion coefficient in three-dimensionally ordered macroporous V_2O_5 films. *Journal of Materials Chemistry C*, 2014. 2(18): p. 3651–3658.

161. Tong, Z., X. Zhang, H. Lv, N. Li, H. Qu, J. Zhao, Y. Li, and X.-Y. Liu. From amorphous macroporous film to 3D crystalline nanorod architecture: A new approach to obtain high-performance V_2O_5 electrochromism. *Advanced Materials Interfaces*, 2015. 2(12): p. 1500230.

162. Lampert, C.M. Electrochromic materials and devices for energy efficient windows. *Solar Energy Materials*, 1984. 11(1): p. 1–27.

163. Li, C.-P., C. Engtrakul, R.C. Tenent, and C.A. Wolden. Scalable synthesis of improved nanocrystalline, mesoporous tungsten oxide films with exceptional electrochromic performance. *Solar Energy Materials and Solar Cells*, 2015. 132: p. 6–14.

164. Wojcik, P.J., L. Santos, L. Pereira, R. Martins, and E. Fortunato. Tailoring nanoscale properties of tungsten oxide for inkjet printed electrochromic devices. *Nanoscale*, 2015. 7(5): p. 1696–1708.

165. Fernandes, M., R. Leones, A.M.S. Costa, M.M. Silva, S. Pereira, J.F. Mano, E. Fortunato, R. Rego, and V. de Zea Bermudez. Electrochromic devices incorporating biohybrid electrolytes doped with a lithium salt, an ionic liquid or a mixture of both. *Electrochimica Acta*, 2015. 161: p. 226–235.

166. Alonso, E., A.M. Sherman, T.J. Wallington, M.P. Everson, F.R. Field, R. Roth, and R.E. Kirchain. Evaluating rare earth element availability: A case with revolutionary demand from clean technologies. *Environmental Science & Technology*, 2012. 46(6): p. 3406–3414.

167. Fernandes, M., R. Leones, S. Pereira, A.M.S. Costa, J.F. Mano, M.M. Silva, E. Fortunato, V. de Zea Bermudez, and R. Rego. Eco-friendly sol-gel derived sodium-based ormolytes for electrochromic devices. *Electrochimica Acta*, 2017. 232: p. 484–494.

168. Cannavale, A., P. Cossari, G.E. Eperon, S. Colella, F. Fiorito, G. Gigli, H.J. Snaith, and A. Listorti. Forthcoming perspectives of photoelectrochromic devices: A critical review. *Energy & Environmental Science*, 2016. 9(9): p. 2682–2719.

169. Cai, G., J. Wang, and P.S. Lee. Next-generation multifunctional electrochromic devices. *Accounts of Chemical Research*, 2016. 49(8): p. 1469–1476.

170. Huang, Y., M. Zhu, Y. Huang, Z. Pei, H. Li, Z. Wang, Q. Xue, and C. Zhi. Multifunctional energy storage and conversion devices. *Advanced Materials*, 2016. 28(38): p. 8344–8364.

171. Bella, F., G. Leftheriotis, G. Griffini, G. Syrrokostas, S. Turri, M. Grätzel, and C. Gerbaldi. A new design paradigm for smart windows: Photocurable polymers for quasi-solid photoelectrochromic devices with excellent long-term stability under real outdoor operating conditions. *Advanced Functional Materials*, 2016. 26(7): p. 1127–1137.

172. Yang, P., P. Sun, Z. Chai, L. Huang, X. Cai, S. Tan, J. Song, and W. Mai. Large-scale fabrication of pseudocapacitive glass windows that combine electrochromism and energy storage. *Angewandte Chemie International Edition*, 2014. 53(44): p. 11935–11939.

173. Yang, P., P. Sun, and W. Mai. Electrochromic energy storage devices. *Materials Today*, 2016. 19(7): p. 394–402.

CHAPTER 3

Recent Advances in Electrochromic Technology

3.1 NEAR-INFRARED EC TECHNOLOGY

The importance of energy efficient smart windows in buildings has been growing over the years. However, the risk associated with the increasing energy use associated with materials production and end-of-life management phases has been a cause of concern. Active and passive technologies, such as renewable energy systems or high-performance insulation materials, have been introduced to reduce the energy and operational cost in buildings. It is worth noting that approximately 60% of the total energy loss of a building comes from its windows. To limit the operational energy losses, thermal and solar transmittance coefficients and air tightness of the windows have been improved in recent years. Multilayer glazing, vacuum glazing, new spacer solutions, aerogels, low emissivity (low-e) coating, glazing cavity gas fills, composite frames, phase change material window products, and smart windows that can change their properties to adjust to outside and indoor conditions are currently-used fenestration systems that have shown to have a large potential for improving window performance. However, while implementing new smart window technologies, manufacturing impacts should be carefully addressed [1–3]. According to a study by Tarantini et al., production processes can impact the environment from 10% to 60% of the overall contribution [4]. Among smart windows, Beatens et al. found that EC technology to be promising. They improve the energy efficiency of the building envelope by reducing thermal loads and electric lighting demand. Compared to a standard passive window with shading, the application of EC technology can reduce daily energy consumption by 20% to 30% [5]. However, due to high cost (varying in a range of 860–1080 $/m^2), the acceptance of EC technology in the window market has been slow. Studies have shown that EC windows will not be cost effective in comparison with passive windows until a price point of 215 $/m^2 can be reached. These studies have also evaluated the cost of a competitive static window (low-SHGC, argon-filled) to be 170 $/m^2 [6–8]. As mentioned in Chapter 2, conventional EC windows are based on active metal oxide layers that are combined with a polymer electrolyte and transparent electrodes deposited on glass surfaces. When a low voltage is applied to the outer transparent conducting electrodes, ions transfer across the ion-conducting layer to the EC layer; the absorption of light by the active layer causes an opaque appearance of the glass. This process is cyclical in nature when reversing the potential, where the ions migrate back, causing a transparent appearance [9,10]. One of the limitations of conventional EC windows is that they cannot harvest sunlight when the window is darkened or tinted, and therefore an EC window limits lighting potential to reduce heat gain on warm days. To address this drawback, a new technology called near-infrared switching electrochromic (NEC) windows has widely been studied and researched [11,12]. In NEC technology, the transmittance of near-infrared (NIR) radiation (about 52% of solar energy) can be tuned without affecting the visible region of the spectrum. Thus, heat can be reduced while letting sunlight pass through the window. As a result, the combined savings of minimizing heat gain while capitalizing on daylighting can be on the order of 10% more than the highest performing EC windows. The best performing sites include offices and midrise residential buildings in climates with hot summers and cold winters, where the energy savings per unit window area range from 50 to 200 kWhm^{-2} per year [12]. NEC windows have a conductive transparent oxide layer (TCO) in the form of metal-oxide nanocrystals, such as indium tin oxide (ITO). These nanocrystals are prepared in solution and applied to a pane of glass or on a flexible substrate. Use of this TCO contrasts with the thin conducting oxide films employed in EC devices. Table 3.1 presents a comparison between traditional EC and next generation NEC devices [3,5,7,8,11,13]. Previous studies have shown that the highest energy consumption in the conventional EC device processing is mainly due to the active EC layer deposition on the glass pane [14,15]. High manufacturing costs are mainly associated with the complex and energy intensive thin film coating techniques like evaporation, sputtering, or chemical vapor deposition (CVD). Posset et al., in their study on the "cradle-to-gate" environmental analysis of electrically switchable shading devices, highlight the significance of substitute energy and resource-intensive processes by less demanding processing methods for the future development of EC windows [16,17]. There is no solid answer as to which manufacturing method will be the most efficient and economical for generating ITO nanocrystal films. Optimization in material utilization, simplifying the production, and reduction of the size and cost of the equipment are some of ways that are being worked out to reduce operational energy and cost. Vacuum deposition of ITO has been identified to be an inefficient technique where only 30% is actually deposited onto the glass. Thus, solution-based deposition techniques can offer better opportunities for reducing industrial indium scrap. Moreover, due to the deficit in pure indium metal supply against demand, the largest production of indium will come from secondary processing techniques. In

TABLE 3.1 Comparison between Conventional EC and NEC Devices

	Conventional EC	NEC
Mode of operation	The transparent or opaque states transmit and block the entire electromagnetic spectrum, including the UV, visible, and near-infrared (NIR) wavelengths	The transparent or opaque states transmit and block only the NIR radiation, allowing the visible wavelengths
Technology	EC film, typically WO_3-layer but also oxides based on W, Mo, Ir, Ti, V, Ni, Nb	EC layer not required: ITO nanocrystals exhibit both conductive and EC properties
Main attributes	In opaque state: • Regulates heating and cooling loads • Reduction of day light levels	In opaque state: • Regulates heating and cooling loads • Reductions in the use of interior lighting
Manufacturing cost	860–1080 \$/m²	To be determined but expected to be lower than conventional EC because of lower manufacturing costs
Switching speed	30 min/m²	<1 min/m²
Durability	High. EC windows by SAGE shows 10^5 cycle (~30 years of life) within the range of −30–60°C	Expected to be better than conventional EC. It was estimated that the charge capacity changed ~4% over the first 2000 cycles
Energy savings	~5.0%	To be determined

(Continued)

TABLE 3.1 (*Continued*) Comparison between Conventional EC and NEC Devices

	Conventional EC	NEC
Manufacturing conditions	High temperature, energy intensive technique	Low-temperature, solution process technique
Aesthetics	Poor aesthetics	Excellent visible transparency
Glare control	Blinds or curtains are not required	Shading needed
Scalable window size	1.6–2.6 m²	To be determined
Energy for switching	0.5 Wh/m²	0.23 Wh/m²
Voltage requirement	0–5.0 V	1.5–4.0 V
Example manufacturers	SAGE Electrochromics, EControl-Glas, and Gesimat	Commercially not available

Source: Jelle, B.P. et al. *Solar Energy Materials and Solar Cells*, 2012. 96: p. 1–23. [3]; Baetens, R. et al. *Solar Energy Materials and Solar Cells*, 2010. 94(2): p. 87–105. [5]; Bechtel, J. et al. *Cleantech to market: Electrochromic coating for dynamic control*. 2011. [7]; Lee, E.S. *Advancement of Electrochromic Windows in: Pft C.E. Commission.* 2006: LBNL. [8]; DeForest, N. et al. *Building and Environment*, 2013. 61: p. 160–168. [11]; Baldassarri, C. et al. *Solar Energy Materials and Solar Cells*, 2016. 156: p. 170–181. [13]

this context, NECs using indium-based compounds gained from the recovery of end-of-use indium seems to be promising. Finally, the indium price has been volatile, and its price is expected to increase over time. Among coating techniques, slot-die coating has been successfully employed in the active layer film deposition for organic photovoltaics (OPV) [17–21]. For this coating system, positive results have been obtained through its environmental assessment. During the development of promising technologies, an early integration of the energy and environmental analysis can help to distinguish strengths from concerns. Thus, before the product enters the market, the designers can make environmentally conscious decisions based on energy and emissions analysis. Despite the promise NEC technologies offer for energy savings, no studies have been reported where the environmental burdens associated with their manufacturing is analyzed [22].

3.2 SELF-POWERED EC TECHNOLOGY

Conventional smart windows consume electricity generated from fossil fuels. To reduce this dependency, renewable resources like solar, wind, geothermal heat, etc., have been exploited to mitigate the energy crisis, environmental pollution, and wastage associated with the excessive consumption of fossil fuels. Solar energy is one of the major renewable energy sources that have been integrated with the smart windows. Self-powered EC windows were demonstrated by Bechinger and team in 1996, thereby introducing the concept of photo-electrochromic cells (PEC). The fabricated PEC device had a configuration similar to dye-sensitized solar cells (DSSC) [23]. In both configurations, dye-sensitized (treated with inorganic dye) TiO_2 nanoparticle film was used as the photoanode. However, WO_3 EC film replaced the Pt counter electrode in the PEC device. Under proper illumination, the photo-electrons generated at the dye-sensitized TiO_2 photoanode transferred to the EC WO_3 electrode through an external circuit, which initiated the intercalation of cations in the electrolyte into the WO_3 crystal lattice to form blue colored $LixWO_3$ compound, resulting in an opaque/darkened device. The colored device could be spontaneously bleached when the illumination was blocked at the short circuit. However, it should be noted that the speed for the bleaching process is rather slow [24,25]. There was another attempt to fabricate a PEC device using Pt as a counter electrode and coating WO_3 to the photoanode. In this case, Pt could accelerate the bleaching process without interfering in the coloration. Wu et al. and team designed a device by integrating solar cells and PEC, resulting in photovoltachromic cells (PVC) which could collect solar energy under illumination and trigger the chromic behavior [26]. The PVC was comprised of a dye-sensitized TiO_2 photoanode and a patterned WO_3/Pt electrode. Compared to PEC, PVC displayed faster bleaching speed at both short circuit (darkened or opaque state) and open circuit (illumination state). Recently, self-powered EC windows constructed using DSSCs, InGaN/GaN solar cells, and perovskite solar cells have been researched and designed. In these self-powered integrated systems, the solar cells convert sunlight into electricity, which is used to power the EC windows [27–31]. The power used by EC windows is then effectively utilized to modulate the throughput of the sunlight and solar heat into the building by reversibly changing the windows' color. Apart from solar energy, mechanical energy can also be used to power smart window systems. For instance, Wang and team members designed nanogenerators to run self-powered smart window systems [32,33]. In their design, a nanogenerator was used to convert ambient mechanical energy from wind and raindrops into electricity and then drive the EC smart windows to change optical properties to colored or bleached state in a reversible manner. In these systems, the maximum optical modulation that can be attained is approximately 32.4% at 695 nm. Tapping chemical energy is another lucrative approach to attain sustainable energy for self-powered smart window systems. Wang et al. in their studies have demonstrated a self-powered EC window which consists of a Prussian blue layer on ITO glass, a strip of Al sheet attached on ITO glass, and 3 mol L^{-1} aqueous KCl as the electrolyte. It was observed that Prussian blue can be transformed to Prussian white (colorless state) by Al sheet in KCl electrolyte resulting from self-bleaching of the window. The device transformed to blue color on disconnecting the Al sheet and Prussian white electrodes. However, the bleached device recovers its pristine blue color through oxidation of oxygen [34]. This switching process requires a long recovery time of 12 h. An addition of trace amounts of strong oxidants in the electrolyte (for example NaClO, H_2O_2, and $(NH_4)2S_2O_8$) can oxidize the Prussian white to Prussian blue within a few minutes. This can speed up the recovery or switching process of the self-powered EC window [35,36].

3.3 MULTIFUNCTIONAL EC SMART WINDOWS

The self-powered EC windows offer various strategies to integrate sustainable energy sources in smart windows. Smart windows can dramatically reduce the energy consumption in buildings by decreasing the cooling loads, heating loads, or electrical lightings. It also offers an attribute of protecting users' privacy. From an operational standpoint, EC windows also behave like energy storage devices with a rational circuit design [37]. The stored energy can be released from EC windows to drive other electronic devices. The EC process is basically related to the intercalation and de-intercalation of ions into and out of the EC electrode. This insertion and extraction process is accompanied by charge separation resulting in energy storage within the electrolyte reservoir. Thus, reutilization of the stored energy in the EC smart window is an important aspect that needs to be exploited in future energy solutions. Smart windows with new design features and functionalities are been widely researched to extend their existing application range, which is of great interest to the scientific community. Multifunctional EC devices incorporating NiO, WO_3, and WO_3/PEDOT:PSS electrodes are widely used in smart windows [38–41]. For an NiO electrode, it was shown that the color of the anodic (NiO) film changes from transparent state to brown during the charging process. The brown color disappears during the reverse discharging process. In this study, the level of stored energy in the NiO film was monitored through the color transition. It was observed that when the NiO electrode was fully charged at a potential of 0.5 V in a KOH electrolyte solution, the film exhibited a dark brown color. On complete discharging, the film recovered its transparent state at a potential of 0 V. The color of WO_3 (cathodic EC) based thin film electrodes changed to deep blue at -0.7 V in H_2SO_4 electrolyte during the charging process. The colored WO_3 film turned to transparent state when the discharge process was completed at a potential of 0 V. Multifunctional smart windows constructed by assembling inkjet-printed WO_3/PEDOT:PSS (cathode) and CeO_2/TiO_2 (anode) films in H_2SO_4 aqueous solution as the electrolyte showed a dark blue colored state at 2.5 V. These colored smart windows could recover the original transparent state by applying a voltage of -2.0 V. They exhibited an optical modulation of about 70% at 633 nm during the charge and discharge process. The level of energy storage could be visually monitored by color changes. It was demonstrated that four such smart windows (each window of dimensions 4.5×4.5 cm^2) could be connected in series and could be used to illuminate a light-emitting diode (LED) for more than 2 h. Other groups have demonstrated similar multifunctional smart windows with excellent performance using multifunctional inorganic and organic materials [27,42,43].

3.4 REFLECTIVE EC SMART WINDOWS

Reflective EC smart windows are types of EC devices that are mainly based on reversible deposition–dissolution of metals like Bi, Ag, Cu, Pb, etc. [44–50]. Reflective smart windows design comprises an EC material dissolved in an electrolyte and fixed between the two parallel transparent electrodes. Reversible reflectance originates from the electrodeposition of the metal layer onto the transparent electrodes and the dissolution of metal into the electrolyte by passing electrical current across the device. During this process, electrodes adjust its optical state. Reflective

EC smart windows have their own advantages and limitations when used in buildings. To reduce interior heating in summer, the light reflecting EC smart window is more effective than that of light absorbing counterparts, as the light absorbing window itself may be heated via the absorption of sunlight. However, light reflecting EC smart windows can cause exterior glare, which is known as light pollution and the visibility of the exterior through the light reflecting EC smart window might be inferior to the light absorbing windows. However, the major challenges come from the electrochemical view point as the light reflecting EC smart windows show poor stability of the mirror state and a lack of bi-stability in the reflectance state, which limits its application in field conditions. Cu-based reversible EC mirror devices that offer reversible switching ability between transparent, blue, and mirror states have been demonstrated. It has been observed that these tristate reversible EC mirror devices can be tuned electrochemically to attain dual transmittance and reflectance modulations using a single device. This offers more options to meet the outdoor and/or indoor glass transition operations [51–53].

3.5 CHALLENGES IN ELECTROLYTES

This book has discussed different types of nanostructured EC materials, preparation processes of large scale EC films, manufacturing considerations, and different kinds of smart windows. From a commercial standpoint many companies including SAGE Electrochromics Inc., ChromoGenics AB, View, Inc., Asahi Glass Co., Ltd., RavenBrick LLC, EControl-Glas, Magna Glass & Window Company, Inc., etc., are actively involved in the research, production, and commercialization of smart windows. However, there are still some challenges that need to be dealt with for commercialization. For instance, apt and low-cost electrolyte systems that can enable fast switching and provide robust electrochemical and environmental stability need further research before they can get implemented. Electrolytes, which are typically used in EC devices, should meet requirements such as high ionic conductivity, high transparency, low volatility, shelf life, electrochemical stability, thermal stability, etc. EC films are being widely researched and tested in aqueous, organic, or ionic liquid (IL)-based electrolytes. Among these liquid electrolytes, IL-based electrolytes have gained more attention due to the wide electrochemical potential window, high ionic conductivity, transference number, and recycling potential [54–58]. For instance, Ho and team members synthesized a thermoplastic solid polymer electrolyte utilizing N,N,N′,N′-tetramethyl-p-phenylenediamine, heptylviologen ($HV(BF_4)_2$), succinonitrile, and poly(vinylidene fluoride-co-hexafluoropropylene) (PVdF-HFP). Different amounts of the IL ($BMIMBF_4$) were added into the solid polymer electrolyte and an EC device was prepared. This EC device exhibited a reversible optical modulation of about 60.1% at 615 nm with a long cycle life [59]. The same team also synthesized an IL electrolyte (1-butyl-3-{2-oxo-2-[(2,2,6,6-tetramethylpiperidin-1-oxyl-4-yl)amino] ethyl}-1H-imidazol-3-ium tetrafluoroborate ($TILBF_4$), containing a stable radical, 2,2,6,6-tetramethyl-1-piperidinyl-oxy for EC application. An EC device was assembled based on $TILBF_4$ and a poly(3,3-diethyl-3,4-dihydro-2H-thieno-[3,4-b][1,4]dioxepine) (PProDOT-Et2) thin film. This EC device exhibited an optical modulation of about 62.2% at 590 nm with fast switching speed of 4.0 and 3.6 s for bleached and coloration processes, respectively, with a coloration efficiency (CE) of 983 $cm^2\ C^{-1}$ at 590 nm. This device retained about 98.0% of its starting optical modulation after 1,000 cycles [60]. Studies by Lu and team proposed a strategy to introduce proton conduction into IL-based electrolytes. The electrolytes were prepared by immersing the sulfonic acid-grafted P(VDF-HFP) electrospun mats in $BMIMBF_4$. These PANI-based EC devices showed high optical modulation about 56.2% at 650 nm) and fast switching speed (2 and 2.5 s for coloration and bleached processes, respectively) [61]. Another study by Shaplov et al. demonstrated an all-polymer-based organic EC device using polymeric ionic liquids as ion conducting separators. The device showed fast switching speed of 3 s, high CE of 390 $cm^2\ C^{-1}$ at

620 nm, and optical modulation up to 22% [62]. Smart windows prepared through copolymerization of N-isopropylacrylamide with (or without) 3-butyl-1-vinyl-imidazolium bromide ionic liquid (here diallylviologen was used as both the cross-linking agent and EC material) showed thermo- and electro-dual responses [63]. However, these liquid electrolytes have certain limitations such as leakage, sealing problems, and solvent evaporation. Gel electrolytes have also been reported, however bubble entrapment during the assembling process and lack of weathering stability are some of the issues that need to be addressed [64,65]. A polymer-based electrolyte with an ionic conductivity of 9.1×10^{-4} S cm^{-1} was obtained through layer-by-layer technology using linear polyethylenimine, poly(ethylene oxide), and poly(acrylic acid) (PAA) and was applied on a flexible ITO/PET substrate [66,67]. The electrolyte exhibited four inter-bonding layers per deposition cycle where electrostatic attraction and hydrogen bonding was primarily responsible for the adhesion. Porous films of [PANI/PAA-PEI]n prepared by self-assembly of a complex polyelectrolyte via the LbL method under an accelerated growth rate showed enhanced EC properties. These films displayed an optical modulation of approximately 30% at 630 nm with fast switching rates. Another multilayer system was reported by alternatively stacking poly(acrylic-acid) (PAA) and polyethylene glycol (PEG)$-\alpha$-cyclodextrin (αCD) complex through LbL self-assembly via hydrogen-bonding [68]. These films with building blocks of PAA and PEG-αCD complex present high ionic conductivity of 2.5×10^{-5} S cm^{-1} at room temperature and relative humidity of 52%. The conductivity value reported in this study was nearly two orders of magnitude higher than PEG/PAA films under similar working conditions. These PAA and PEG-αCD complex electrolytes provide solutions for designing practical polymer-based solid-state electrolytes for EC windows. Improvements in the EC performances are rather challenging, often resulting in trade-off between high energy storage density, fast switching speed, and large optical modulation in multifunctional energy storage smart windows [69,70].

3.6 EPILOGUE

Packaging of smart windows is a technological challenge, and very little literature is available on this topic. Sealants that are chemically stable are an important requirement for obtaining long life EC devices. In the reported studies, acetic silicone and DuPont's thermoplastic Surlyn were claimed to be used as sealants in the assembled EC device [69,70]. Cost is another important factor that has restricted market infiltration of smart windows. The current cost of EC windows in the commercial market is $538 to $1076 per m² as estimated by National Renewable Energy Laboratory (NREL). This high cost has prevented widespread acceptance and acknowledgment of the EC technology in the industry. As per NREL's estimation, if the price of smart windows can be reduced to $200 per m², then it will be an attractive and cost-competitive option for residential building applications [71]. In addition to this, more efforts are required to explore multifunctional smart windows that can provide value added functionalities without affecting the performance. This will help in integrating EC devices into green building systems. Finally, it must be mentioned that success of any EC configuration will depend on new materials (micro/nano) used, large scale manufacturing strategies, and rational device designs that will provide solutions to address these challenges and promote a rapid growth for both research and commercialization of multifunctional smart windows.

REFERENCES

1. Asdrubali, F., C. Baldassarri, and V. Fthenakis. Life cycle analysis in the construction sector: Guiding the optimization of conventional Italian buildings. *Energy and Buildings*, 2013. 64: p. 73–89.
2. Asdrubali, F., and G. Baldinelli. Theoretical modelling and experimental evaluation of the optical properties of glazing systems with selective films. *Building Simulation*, 2009. 2(2): p. 75–84.
3. Jelle, B.P., A. Hynd, A. Gustavsen, D. Arasteh, H. Goudey, and R. Hart. Fenestration of today and tomorrow: A state-of-the-art review and future research opportunities. *Solar Energy Materials and Solar Cells*, 2012. 96: p. 1–28.

4. Tarantini, M., A.D. Loprieno, and P.L. Porta. A life cycle approach to green public procurement of building materials and elements: A case study on windows. *Energy*, 2011. 36(5): p. 2473–2482.

5. Baetens, R., B.P. Jelle, and A. Gustavsen. Properties, requirements and possibilities of smart windows for dynamic daylight and solar energy control in buildings: A state-of-the-art review. *Solar Energy Materials and Solar Cells*, 2010. 94(2): p. 87–105.

6. Gillaspie, D.T., R.C. Tenent, and A.C. Dillon. Metal-oxide films for electrochromic applications: Present technology and future directions. *Journal of Materials Chemistry*, 2010. 20(43): p. 9585–9592.

7. Bechtel, J., K. Isono, and S. Lounis. *Cleantech to market: Electrochromic coating for dynamic control.* 2011.

8. Lee, E.S. *Advancement of Electrochromic Windows in: Pft C.E. Commission.* 2006: LBNL.

9. Granqvist, C.G. Electrochromic oxides: A bandstructure approach. *Solar Energy Materials and Solar Cells*, 1994. 32(4): p. 369–382.

10. Lee, E.S., S.E. Selkowitz, R.D. Clear, D.L. DiBartolomeo, J.H. Klems, L.L. Fernandes, G.J. Ward, and V. Inkarojrit. *A Design Guide for Early-Market Electrochromic Windows.* 2006, Berkeley.

11. DeForest, N., A. Shehabi, G. Garcia, J. Greenblatt, E. Masanet, E.S. Lee, S. Selkowitz, and D.J. Milliron. Regional performance targets for transparent near-infrared switching electrochromic window glazings. *Building and Environment*, 2013. 61: p. 160–168.

12. DeForest, N., A. Shehabi, J. O'Donnell, G. Garcia, J. Greenblatt, E.S. Lee, S. Selkowitz, and D.J. Milliron. United States energy and CO_2 savings potential from deployment of near-infrared electrochromic window glazings. *Building and Environment*, 2015. 89: p. 107–117.

13. Baldassarri, C., A. Shehabi, F. Asdrubali, and E. Masanet. Energy and emissions analysis of next generation electrochromic devices. *Solar Energy Materials and Solar Cells*, 2016. 156: p. 170–181.

14. Papaefthimiou, S., G. Leftheriotis, and P. Yianoulis. Advanced electrochromic devices based on WO_3 thin films. *Electrochimica Acta*, 2001. 46(13): p. 2145–2150.

15. Papaefthimiou, S., E. Syrrakou, and P. Yianoulis. Energy performance assessment of an electrochromic window. *Thin Solid Films*, 2006. 502(1): p. 257–264.

16. Posset, U., M. Harsch, A. Rougier, B. Herbig, G. Schottner, and G. Sextl. Environmental assessment of electrically controlled variable light transmittance devices. *RSC Advances*, 2012. 2(14): p. 5990–5996.

17. Krebs, F.C. Fabrication and processing of polymer solar cells: A review of printing and coating techniques. *Solar Energy Materials and Solar Cells*, 2009. 93(4): p. 394–412.

18. Espinosa, N., R. García-Valverde, A. Urbina, F. Lenzmann, M. Manceau, D. Angmo, and F.C. Krebs. Life cycle assessment of ITO-free flexible polymer solar cells prepared by roll-to-roll coating and printing. *Solar Energy Materials and Solar Cells*, 2012. 97: p. 3–13.

19. García-Valverde, R., J.A. Cherni, and A. Urbina. Life cycle analysis of organic photovoltaic technologies. *Progress in Photovoltaics: Research and Applications*, 2010. 18(7): p. 535–558.

20. Espinosa, N., R. García-Valverde, A. Urbina, and F.C. Krebs. A life cycle analysis of polymer solar cell modules prepared using roll-to-roll methods under ambient conditions. *Solar Energy Materials and Solar Cells*, 2011. 95(5): p. 1293–1302.

21. Llordés, A., G. Garcia, J. Gazquez, and D.J. Milliron. Tunable near-infrared and visible-light transmittance in nanocrystal-in-glass composites. *Nature*, 2013. 500: p. 323.

22. Baldassarri, C., F. Mathieux, F. Ardente, C. Wehmann, and K. Deese. Integration of environmental aspects into R&D inter-organizational projects management: Application of a life cycle-based method to the development of innovative windows. *Journal of Cleaner Production*, 2016. 112: p. 3388–3401.

23. Bechinger, C., S. Ferrere, A. Zaban, J. Sprague, and B.A. Gregg. Photoelectrochromic windows and displays. *Nature*, 1996. 383: p. 608.

24. Krašovec, U.O., A. Georg, A. Georg, V. Wittwer, J. Luther, and M. Topič. Performance of a solid-state photoelectrochromic device. *Solar Energy Materials and Solar Cells*, 2004. 84(1): p. 369–380.

25. Georg, A., A. Georg, and U. Opara Krašovec. Photoelectrochromic window with Pt catalyst. *Thin Solid Films*, 2006. 502(1): p. 246–251.

26. Wu, J.-J., M.-D. Hsieh, W.-P. Liao, W.-T. Wu, and J.-S. Chen. Fast-switching photovoltachromic cells with tunable transmittance. *ACS Nano*, 2009. 3(8): p. 2297–2303.

27. Xie, Z., X. Jin, G. Chen, J. Xu, D. Chen, and G. Shen. Integrated smart electrochromic windows for energy saving and storage applications. *Chemical Communications*, 2014. 50(5): p. 608–610.

28. Wu, C.-C., J.-C. Liou, and C.-C. Diao. Self-powered smart window controlled by a high open-circuit voltage InGaN/GaN multiple quantum well solar cell. *Chemical Communications*, 2015. 51(63): p. 12625–12628.

29. Xia, X., Z. Ku, D. Zhou, Y. Zhong, Y. Zhang, Y. Wang, M.J. Huang, J. Tu, and H.J. Fan. Perovskite solar cell powered electrochromic batteries for smart windows. *Materials Horizons*, 2016. 3(6): p. 588–595.

30. Cannavale, A., G.E. Eperon, P. Cossari, A. Abate, H.J. Snaith, and G. Gigli. Perovskite photovoltachromic cells for building integration. *Energy & Environmental Science*, 2015. 8(5): p. 1578–1584.

31. Zhou, F., Z. Ren, Y. Zhao, X. Shen, A. Wang, Y.Y. Li, C. Surya, and Y. Chai. Perovskite photovoltachromic supercapacitor with all-transparent electrodes. *ACS Nano*, 2016. 10(6): p. 5900–5908.

32. Yang, X., G. Zhu, S. Wang, R. Zhang, L. Lin, W. Wu, and Z.L. Wang. A self-powered electrochromic device driven by a nanogenerator. *Energy & Environmental Science*, 2012. 5(11): p. 9462–9466.

33. Yeh, M.-H., L. Lin, P.-K. Yang, and Z.L. Wang. Motion-driven electrochromic reactions for self-powered smart window system. *ACS Nano*, 2015. 9(5): p. 4757–4765.

34. Wang, J., L. Zhang, L. Yu, Z. Jiao, H. Xie, X.W. Lou, and X. Wei Sun. A bi-functional device for self-powered electrochromic window and self-rechargeable transparent battery applications. *Nature Communications*, 2014. 5: p. 4921.

35. Zhang, H., Y. Yu, L. Zhang, Y. Zhai, and S. Dong. Self-powered fluorescence display devices based on a fast self-charging/recharging battery (Mg/Prussian blue). *Chemical Science*, 2016. 7(11): p. 6721–6727.

36. Zhao, J., Y. Tian, Z. Wang, S. Cong, D. Zhou, Q. Zhang, M. Yang, W. Zhang, F. Geng, and Z. Zhao. Trace H_2O_2-assisted high-capacity tungsten oxide electrochromic batteries with ultrafast charging in seconds. *Angewandte Chemie International Edition*, 2016. 55(25): p. 7161–7165.

37. Kumar, V., S. Park, K. Parida, V. Bhavanasi, and P.S. Lee. Multi-responsive supercapacitors: Smart solution to store electrical energy. *Materials Today Energy*, 2017. 4: p. 41–57.

38. Cai, G., P. Darmawan, X. Cheng, and P.S. Lee, Inkjet printed large area multifunctional smart windows. *Advanced Energy Materials*, 2017. 7(14): p. 1602598.

39. Cai, G., P. Darmawan, M. Cui, J. Wang, J. Chen, S. Magdassi, and P.S. Lee. Supercapacitors: Highly stable transparent conductive silver grid/PEDOT:PSS electrodes for integrated bifunctional flexible electrochromic supercapacitors (Adv. Energy Mater. 4/2016). *Advanced Energy Materials*, 2016. 6(4).

40. Cai, G., X. Wang, M. Cui, P. Darmawan, J. Wang, A.L.-S. Eh, and P.S. Lee. Electrochromo-supercapacitor based on direct growth of NiO nanoparticles. *Nano Energy*, 2015. 12: p. 258–267.

41. Cai, G., J. Wang, and P.S. Lee. Next-generation multifunctional electrochromic devices. *Accounts of Chemical Research*, 2016. 49(8): p. 1469–1476.

42. Wang, K., H. Wu, Y. Meng, Y. Zhang, and Z. Wei. Integrated energy storage and electrochromic function in one flexible device: An energy storage smart window. *Energy & Environmental Science*, 2012. 5(8): p. 8384–8389.

43. Yang, P., P. Sun, Z. Chai, L. Huang, X. Cai, S. Tan, J. Song, and W. Mai. Large-Scale fabrication of pseudocapacitive glass windows that combine electrochromism and energy storage. *Angewandte Chemie International Edition*, 2014. 53(44): p. 11935–11939.

44. Oliveira, M.R.S., D.A.A. Mello, E.A. Ponzio, and S.C. de Oliveira. KI effects on the reversible electrodeposition of silver on poly(ethylene oxide) for application in electrochromic devices. *Electrochimica Acta*, 2010. 55(11): p. 3756–3765.

45. de Mello, D.A.A., M.R.S. Oliveira, L.C.S. de Oliveira, and S.C. de Oliveira. Solid electrolytes for electrochromic devices based on reversible metal electrodeposition. *Solar Energy Materials and Solar Cells*, 2012. 103: p. 17–24.

46. Nakashima, M., T. Ebine, M. Shishikura, K. Hoshino, K. Kawai, and K. Hatsusaka. Bismuth electrochromic device with high paper-like quality and high performances. *ACS Applied Materials & Interfaces*, 2010. 2(5): p. 1471–1482.

47. Richardson, T.J. New electrochromic mirror systems. *Solid State Ionics*, 2003. 165(1): p. 305–308.

48. Avellaneda, C.O., M.A. Napolitano, E.K. Kaibara, and L.O.S. Bulhões. Electrodeposition of lead on ITO electrode: Influence of copper as an additive. *Electrochimica Acta*, 2005. 50(6): p. 1317–1321.

49. Howard, B.M., and J.P. Ziegler. Optical properties of reversible electrodeposition electrochromic materials. *Solar Energy Materials and Solar Cells*, 1995. 39(2): p. 309–316.

50. Jeong, K.R., I. Lee, J.Y. Park, C.S. Choi, S.-H. Cho, and J.L. Lee. Enhanced black state induced by spatial silver nanoparticles in an electrochromic device. *Npg Asia Materials*, 2017. 9: p. e362.

51. Araki, S., K. Nakamura, K. Kobayashi, A. Tsuboi, and N. Kobayashi. Electrochemical optical-modulation device with reversible transformation between transparent, mirror, and black. *Advanced Materials*, 2012. 24(23): p. OP122–OP126.

52. Park, C., S. Seo, H. Shin, B.D. Sarwade, J. Na, and E. Kim. Switchable silver mirrors with long memory effects. *Chemical Science*, 2015. 6(1): p. 596–602.

53. Eh, A.L.-S., M.-F. Lin, M. Cui, G. Cai, and P.S. Lee. A copper-based reversible electrochemical mirror device with switchability between transparent, blue, and mirror states. *Journal of Materials Chemistry C*, 2017. 5(26): p. 6547–6554.

54. Zhou, D., R. Zhou, C. Chen, W.A. Yee, J. Kong, G. Ding, and X. Lu. Non-Volatile polymer electrolyte based on poly(propylene carbonate), ionic liquid, and lithium perchlorate for electrochromic devices. *The Journal of Physical Chemistry B*, 2013. 117(25): p. 7783–7789.

55. Cruz, H., N. Jordão, and L.C. Branco. Deep eutectic solvents (DESs) as low-cost and green electrolytes for electrochromic devices. *Green Chemistry*, 2017. 19(7): p. 1653–1658.

56. Wei, Y., J. Zhou, J. Zheng, and C. Xu. Improved stability of electrochromic devices using Ti-doped V_2O_5 film. *Electrochimica Acta*, 2015. 166: p. 277–284.

57. Zhang, J., J.P. Tu, G.H. Du, Z.M. Dong, Y.S. Wu, L. Chang, D. Xie, G.F. Cai, and X.L. Wang. Ultra-thin WO_3 nanorod embedded polyaniline composite thin film: Synthesis and electrochromic characteristics. *Solar Energy Materials and Solar Cells*, 2013. 114: p. 31–37.

58. Thakur, V.K., G. Ding, J. Ma, P.S. Lee, and X. Lu. Hybrid materials and polymer electrolytes for electrochromic device applications. *Advanced Materials*, 2012. 24(30): p. 4071–4096.

59. Chang, T.H., C.W. Hu, S.Y. Kao, C.W. Kung, H.W. Chen, and K.C. Ho. An all-organic solid-state electrochromic device containing poly(vinylidene fluoride-co-hexafluoropropylene), succinonitrile, and ionic liquid. *Solar Energy Materials and Solar Cells*, 2015. 143: p. 606–612.

60. Fan, M.S., C.P. Lee, R. Vittal, and K.C. Ho. A novel ionic liquid with stable radical as the electrolyte for hybrid type electrochromic devices. *Solar Energy Materials and Solar Cells*, 2017. 166: p. 61–68.

61. Zhou, R., W. Liu, Y.W. Leong, J. Xu, and X. Lu. Sulfonic acid- and lithium sulfonate-grafted poly(vinylidene fluoride) electrospun mats as ionic liquid host for electrochromic device and lithium-ion battery. *ACS Applied Materials & Interfaces*, 2015. 7(30): p. 16548–16557.

62. Shaplov, A.S., D.O. Ponkratov, P.-H. Aubert, E.I. Lozinskaya, C. Plesse, F. Vidal, and Y.S. Vygodskii. A first truly all-solid state organic electrochromic device based on polymeric ionic liquids. *Chemical Communications*, 2014. 50(24): p. 3191–3193.

63. Chen, F., Y. Ren, J. Guo, and F. Yan. Thermo- and electro-dual responsive poly(ionic liquid) electrolyte based smart windows. *Chemical Communications*, 2017. 53(10): p. 1595–1598.

64. Agnihotry, S.A., P. Pradeep, and S.S. Sekhon. PMMA based gel electrolyte for EC smart windows. *Electrochimica Acta*, 1999. 44(18): p. 3121–3126.

65. Souza, F.L., M.A. Aegerter, and E.R. Leite. Performance of a single-phase hybrid and nanocomposite polyelectrolyte in classical electrochromic devices. *Electrochimica Acta*, 2007. 53(4): p. 1635–1642.

66. Nguyen, C.A., S. Xiong, J. Ma, X. Lu, and P.S. Lee. High ionic conductivity P(VDF-TrFE)/PEO blended polymer electrolytes for solid electrochromic devices. *Physical Chemistry Chemical Physics*, 2011. 13(29): p. 13319–13326.

67. Nguyen, C.A., A.A. Argun, P.T. Hammond, X. Lu, and P.S. Lee. Layer-by-layer assembled solid polymer electrolyte for electrochromic devices. *Chemistry of Materials*, 2011. 23(8): p. 2142–2149.

68. Cui, M., and P.S. Lee. Solid polymer electrolyte with high ionic Conductivity via layer-by-layer deposition. *Chemistry of Materials*, 2016. 28(9): p. 2934–2940.

69. Syrrakou, E., S. Papaefthimiou, and P. Yianoulis. Environmental assessment of electrochromic glazing production. *Solar Energy Materials and Solar Cells*, 2005. 85(2): p. 205–240.

70. Byker, H.J. Electrochromics and polymers. *Electrochimica Acta*, 2001. 46(13): p. 2015–2022.

71. Roberts, D.R. *Preliminary Assessment of the Energy-Saving Potential of Electrochromic Windows in Residential Buildings*. National Renewable Energy Laboratory Technical Report, 2006. TP-550-46916: p. 1–12.

Index

Printed and bound by CPI Group (UK) Ltd, Croydon, CR0 4YY

24/10/2024

01778283-0001